8-5-76

An introduction to structural design

Timber

Books by the same author, Derrick Beckett:
An introduction to Structural design (1) Concrete Bridges
Limit State Design of Reinforced Concrete Structures

An introduction to structural design
Timber

DERRICK BECKETT AND PAUL MARSH

SURREY UNIVERSITY PRESS
IN ASSOCIATION WITH
INTERTEXT PUBLISHING LIMITED

Surrey University Press
Kingswood House, Heath & Reach, nr Leighton Buzzard, Beds LU7 OAZ
and 450 Edgware Rd., London W2 1EG
© Derrick Beckett and Paul Marsh 1974

First published 1974

ISBN 0 903384 02 7

 PRINTED BY Unwin Brothers Limited
THE GRESHAM PRESS OLD WOKING SURREY ENGLAND

Produced by 'Uneoprint'·

A member of the Staples Printing Group

OF TIMBER

'Timber (Vitruvius hath it, cap. 9, book 2) ought to be felled in Autumn, and through all the Winter: because then the Trees recover from the Root that strength and soundness which in the Spring and Summer was dispersed into leaves and fruit; and you are to cut them in the wane of the Moon, because the moisture which is most apt to rot wood, is then consumed: From whence there will not come the worm to hurt it. It should be cut but to the middle of the pith, and so left until it be dry, because by drops there will pass away that moisture which would cause putrefication; being cut, let it be laid in a place free from the extremity of the Sun, Wind and Rain; and those ought chiefly to be kept dry, which are of spontaneous growth; and to the end that they may not cleave but dry equally, you are to daub them over with Cow-dung; it should not be drawn through the dew, but in the afternoon, nor to be wrought being very wet or too dry: Because the one makes it apt to rot, the other hard to work; nor will it in less than Three Years be dry enough, to use in Plankes, Doors and Windows. It is convenient for those who are about to build, to inform themselves from men skilful in the nature of Timber what Wood is fit for such use, and what not. Vitruvius, in the Chapter above mentioned gives good instructions; and so other learned men who have written thereof at large.'

The First Book of Architecture   ANDREA PALLADIO (1518-1580)

Preface

This book is an attempt to illustrate that timber—one of man's oldest building materials—is particularly relevant to contemporary building. It has been a material often overlooked, due to architectural and engineering concentration on the so-called modern structural materials—steel and concrete. It has also been ignored due to allegations of lack of durability and fire resistance.

The authors have set out, against the background of afforestation policies and environmental considerations, to examine timber's use in building through the ages and the recent developments which have overcome previous limitations of dimension and durability and have allowed timber to be considered as a viable structural material. They have proceeded to demonstrate the particular relevance of timber to industrialised building.

The co-authors, Derrick Beckett and Paul Marsh, have written from their respective backgrounds of engineering and architecture, drawing upon their collaboration in the development of commercial system buildings in timber.

Contents

Acknowledgements

Acknowledgement must be made to the following for their help and advice during the preparation of this book:

D P Rooke, Principal Information Officer, Forestry Commission

Arne Jacobsen of Hellerup, Denmark for details and illustration of the Kubeflex system (Fig. 12/18)

Richard Saunders Allen of New York for Information on American trussed bridge construction (Fig. 2/16)

Members of the staff of the Timber Research and Development Association

Building Research Establishment, Princes Risborough Laboratory, Princes Risborough, Aylesbury, Buckinghamshire, Chapter 4 data for table 4.1 and Fig. 4/6: Chapter 5 data for tables 5.1, 5.2, 5.3, 5.4: Chapter 6, Fig. 6/4: Chapter 7, table 7.1, Chapter 9, Fig. 9/39.

Reference has been made in the text to publications of the Forest Products Research Laboratory (Timberlab) and it should be noted that FPRL is now the Princes Risborough Laboratory (PRL) of the Building Research Establishment, but is still to be found on the same site.

British Standards Institution: The tables and data reproduced from CP112: Part 2: 1971 Structural use of timber in Chapters 5, 8, 9 and 15 are reproduced by permission of the British Standards Institution 2, Park Street, London W1A 2BS. This code of practice is now under revision.

The Swedish Timber Council, Sweden House, 14, Trinity Square, London EC3N 4BN. Data on the shrinkage of Redwood and Whitewood given in Chapter 6: Chapter 7, Fig. 7/1.

E & F N Spon Ltd., 11 New Fetter Lane, London EC4. Permission to reproduce data from The Structural Use of Timber, a commentary on the British Standard Code of Practice CP112 by L G Booth and P O Reece, 1967. Permission to reproduce data given in Chapter 9 relating to nails, screws and bolts also Figs. 9/26, 9/28, 9/29 and 9/30.

Acknowledgement of permission to reproduce illustrations from the following bodies:

Western Region of British Rail (Fig. 2/14)

Council of Forest Industries of British Columbia (Figs. 3/2, 3/3 and 3/4)

Controller of Her Majesty's Stationery Office (Figs. 1/7, 12/3, 12/7 Crown Copyright)

Forestry Commission (Figs. 1/1, 1/3 and 1/4)

SGB Group of Companies (Peter Cox Ltd) (Figs. 10/1, 10/3 and 10/4)

Oxford Regional Hospital Board and CED Building Services of British Steel Corporation (Figs. 12/11)

A S Byggeselskabet (Fig. 12/17)

CIBA-GEIGY Plastics Division (Figs. 11/1 and 11/3)

Aluminium Federation (Figs. 12/4, 12/5 and 12/6)

Rainham Timber Engineering Co Ltd (Figs. 3/10, 3/11 and 9/15)

Foster Construction Services (Fig. 12/16)

Protimeter Limited, Fieldhouse Lane, Marlow, Bucks SL7 1LS. Illustrations of the Protimeter Timbermaster and the Hammer Electrode, (Figs. 6/2 and 6/3.)

Macandrews and Forbes Ltd., Pembroke House, 44 Wellesley Road, Croydon CR9 3QE (Figs. 9/36 and 9/37)

Automatic Pressings Ltd (Bat Products) Halesfield Industrial Estate, Telford, Shropshire TF7 4LD (Fig. 9/38)

Automated Building Components (UK) Ltd., The Trading Estate, Farnham, Surrey (Figs. 9/32 and 9/33)

1 Why wood?

Why should anyone in this time of increasing sophistication and technical progress, with exciting new materials presenting themselves as possible successors to traditional natural building materials, still seriously consider the use of timber? Can this most traditional of materials still hold its place along side other, newer materials, or should it be relegated to the position of a second class building material?

Those very advantages that singled timber out as the first building material and the one most universally used, have led to an ever-increasing demand for wood for a multitude of purposes, not the least of which are associated with building. Apart from timber's natural advantage of an attractive appearance which can enhance buildings in very many ways, the more functional advantages can be stated as follows: *availability; cheapness; strength; workability; capability of accepting new techniques; compatability with the principles of pre-fabrication.*

AVAILABILITY

Of the earth's total land surface, 22.5 per cent is covered with forest. This is approximately 3, 000 million hectares. These forests vary from the coniferous forests of softwoods (Fig. 1.1)—pines, spruces, firs, cedars, larches, cypresses etc., stretching across the cold temperate zone of the Northern Hemisphere from North America, through Scandinavia to Siberia, to the tropical hardwood forests which are barely yet touched as a timber supply source and consist of numberless deciduous species (balsa, boxwood, ebony, crocuswood, lignum vitae, mahogony, rosewood, satinwood, teak, etc.) Between these extremes there are the temperate broad-leafed hardwood forests (Fig. 1.2) made up of species such as oak, maple, walnut, ash, beech etc.

The resources are obviously enormous, in spite of felling that has taken place throughout the years to meet the increasing demand for timber and to provide space for agricultural purposes or building development. Of the forests in France, Spain, Belgium, Italy and Greece only about 15 per cent remain, in Sweden and Finland about 60 per cent; while in Britain only 5 per cent still exists.

In this country deforestation has been more devastating than in any other European country in spite of early concern about the protection of forests which prompted Edward IV in 1482 to pass an 'Act for Enclosing of Woods'. In 1543 Henry VIII passed a further 'Act for the Preservation of Woods'. James IV of Scotland in 1504 required every landowner to establish at least one acre of wood 'where there are no great woods or forests'. But these efforts were not effective enough to save the dissipation of Britain's natural timber resources. By the Industrial Revolution, home-grown timber sources could not cope with

1

Fig.1.1 A stand of softwoods. (Copyright: Forestry Commission)

*Fig.1.2 Beeches growing in the Chilterns, at Hampden.
(Copyright: Forestry Commission)*

the enlarged requirement, but the enormous resources of North America and Europe provided apparently limitless cheap imported timber. It is for this reason that most timber utilising industries, such as wood pulping mills, were established near ports and not near areas of natural forests.

Britain now imports more timber or timber products than any other country in the world, while its own resources only succeed in barely satisfying 10 per cent of its overall requirements and only $2\frac{1}{2}$ per cent of softwood sawlogs for conversion into structural timbers. (Sawlogs can be defined by minimum dimension. They are logs of a diameter of at least 200 mm and a length of 3 m, these being the smallest sizes for economic conversion). As Britain has little mature timber, the largest proportion of annual output from our newly planted forests is small material used for pulping, mining timbers, poles and stakes, and chipboard manufacture.

In 1919 the British Government, shaken into realisation of the situation by World War I, finally decided to do something about the home timber supply position. In the Forestry Act of that year the Forestry Commission was established to control and manage an afforestation policy. In 1943 a 50 year programme was drawn up whose aim was the establishment of 2 million hectares of forest, or 10 per cent of the total land surface. This, it was estimated, would yield in time, 11 million cubic metres of timber each year, or one third of the country's estimated timber requirement. Between 1967 and 1976 the Forestry Commission aims to plant 240, 000 hectares concentrating on the agriculturally poor upland areas of Scotland and Wales. In 1971 the Commission held 1, 210, 000 hectares of land; 730, 000 hectares under tree crops. That year about 23, 400 hectares were planted with a total of 77 million trees, mostly sitka spruce (Pica sitchensis Carr). The Commission acquires approximately 12, 000 hectares of land per year and, in addition to growing and marketing its timber crops, sponsors and carries out extensive research programmes.

As well as Forestry Commission forests, private woodlands (many receiving Forestry Commission assistance in advice, finance and schemes of planned management) in 1970 covered 1, 112, 000 hectares and 22, 000 hectares were planted.

The expected output of softwood from Commission and privately owned woodlands is set out in graph (Fig. 1. 3). In the twenty year period it is anticipated that output will increase by 70 per cent from private sources and 230 per cent from the Commission.

In addition to the spin-off advantages of this afforestation—that of improving the environment and creating recreation areas, the Forestry Commission is already making a significant inroad into the timber requirements of the country. Bearing in mind that softwoods usually require about 50 years to mature, the majority of the produce at the moment is in thinnings. The Commission has concentrated on softwoods because of their speed of maturing and their greater use for structural

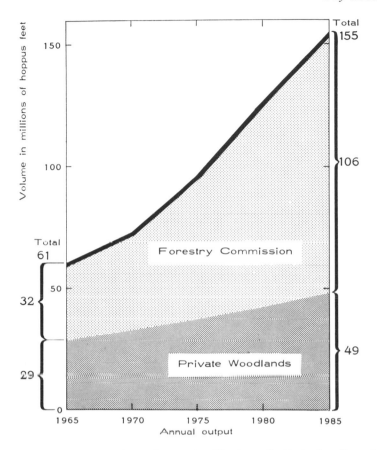

Fig.1.3 How softwood output will rise. Output of softwood from privately-owned woodlands will increase by 70 per cent over the twenty years 1965-1985, from 29 to 48 million hoppus feet per annum. Over the same period, the Forestry Commission output will go up by 230 per cent, from 32 to 106 million. A combined increase from 61 to 155 million hoppus feet, or 154 per cent on the 1965 figure.
The output of hardwood, nearly all of it from private estates, is expected to remain stationary at about 50 million hoppus feet per annum, and is therefore not illustrated. (Copyright: Forestry Commission)

purposes, chipboard manufacture and pulping. In the 1970/71 financial year the Forestry Commission made £4.36 million income from timber sales and 1.57 million cu.m. of timber were felled. During that year timber imports stood at £200 million. Most of this was in Baltic redwood, Baltic whitewood (Spruce), Western Red Cedar, Hemlock and Douglas Fir, nearly all of which will be able to be matched by UK produce as more plantations reach maturity.

And so timber has been recognised (albeit rather late) as a national asset and something is being done to reinstate the forests that have been previously destroyed. At the same time the amenity of the country-side is being improved. This recognition of the need for afforestation can be seen in other parts of the world. Even some deserts in Africa are being reclaimed by careful and staged planting in which the growing of trees forms a vital part. It must be remembered that trees are valuable in their own right as they not only combat the build-up of carbon dioxide in the atmosphere but also are one of the main sources of replenishment of the atmosphere's oxygen and water.

Plato understood the importance of trees. In his 'Critias' he cites the hills of Attica which had become sterile due to indiscriminate felling of trees and the consequent erosion of the soil that followed. The great floods of the Arno, that periodically cover Florence in mud, have been experienced since the fourteenth century when the woodlands were cut down and the sheep and goats that grazed on the grass remaining nibbled it so closely that it died and the area became a baked sterile waste. Trees have got plenty of desirable characteristics in addition to their wood, of which flood prevention and air conditioning are two of the less obvious.

Still the resources of the world's vast forests make timber a material that is readily available, and one which will continue so indefinitely provided felling is not indiscriminate and an effort is made to afforestate and tend our woodlands properly.

CHEAPNESS

With shipping costs rising steeply, timber is no longer the very cheap material it once was, but compared with other structural materials such as steel and concrete, it is still cheap. However expensive timber becomes as a material, its ease of handling, transport and working will always make it compare favourably with steel, and with rising labour costs it is reasonable to expect timber to remain at very competitive prices. The likelihood of an increasing volume of home-grown timber on the market and greater mechanisation in forestry and timber conservation will also help to maintain the situation.

The greater sophistication which has recently been brought into timber engineering has led to more economic use of wood—a reduction in the size of timber members to achieve the same effect and a utilisation of timber waste. This has been possible by a more scientific appraisal of timber as a structural material and the development of synthetic resin glues. Wood takes a long time to grow; it must not be squandered thoughtlessly.

STRENGTH

Timber's main constituent is cellulose—and cellulose is nature's strong material. It is a polymer consisting of long chainlike molecules made of repeated linkings of small molecules, or monomers. If depoly- merised, wood can be turned into sugars. The chain molecules are rigid and more or less parallel which gives wood good strength in one direc- tion. Wood is expanded cellulose impregnated with a resinous material called lignin. The different species of timber are made of the same material expanded to different extents. For instance, Balsa with a specific gravity of 0.10 has 94 per cent of its volume taken up by pores; a very dense concentrated wood (Lignastone) has only 15 per cent of its volume taken up by pores.

Wood is a light material with a high bending strength. Its strength to weight ratio is excellent and this characteristic results in light struc- tures. For many years it was the only material available for spanning large open spaces without resorting to the cumbersome business of stone vaulting. But even today, it is still a worthy alternative to steel or concrete, using the new engineering techniques available.

Timber has one further advantage that has a bearing on its structural capacity. It has a very low co-efficient of thermal expansion—approx- imately 0.0000038 mm per °C. For all practical purposes it can be said to be dimensionally unaffected by temperature changes.

Because of the lightness of wooden structures, the cost of surrounding elements such as foundations, is minimised. But this lightness also increases the vulnerability to up-lift effects of wind pressure. The engineering aspects of wood are investigated more fully in a later chapter.

WORKABILITY

Timber is easy to work with essentially simple tools, many of which have scarcely changed for several hundred years. Its very tradition is at once its strength and also its weakness. We know so much about timber that we are inclined to overlook its very special characteristics.

Wood is essentially a versatile material, capable of intricate fabrication using simple machines, and inexpensive jigs. Timber can be connected in a variety of simple ways using adhesives, nails, screws or bolts. Even large timber assemblies are light and can be handled by manpow- er or the simplest of lifting devices.

Compare all this with the complexities of equipment and fixing devices required for steelwork, or the weight of structural elements in precast concrete.

CAPABILITY OF ACCEPTING NEW TECHNIQUES

Wood has great development potential. From the rough hewn timbers, little better than part tree trunks, used in the Middle Ages (Fig. 1.4)

Fig.1.4 Laverham Cottages.

timber buildings through the years have developed into refined frame-works (Fig. 1. 5). With the development of veneer cutting processes and the improvement in glue technology, a whole new range of possibilities was opened up based on the use of plywood.

Timber in spite of being so commonly used and therefore often under-estimated in potential, is in fact a sophisticated material to which only relatively recently have engineering principles been applied. These have had their most exciting expression in laminated structures, which place timber beside steel, concrete and plastics, as a twentieth century material in its own right (Fig. 1. 6).

In fact timber today is potentially an entirely different structural mater-ial from the one that was known fifty years ago.

Fig.1.5 Interior without lining

Fig.1.6 Footbridge constructed in laminated timber. Crown Copyright

COMPATIBILITY WITH PRINCIPLES OF PREFABRICATION

More and more of the building process now takes place, not on the site, but in the factory; whether it be in small prefabricated assemblies like standard windows and door assemblies, or whole building systems. Timber, because of its lightness, ease of fabrication and transportation, is sympathetic to the principle of prefabrication. Large assemblies can be made in the factory with the care and accuracy that factory conditions will allow. They can be handled there with the simplest of equipment, even when whole sections of building are concerned. In fact only the limitation of transportation widths often determine the eventual sizes of timber assemblies delivered to site. Here they are handled into position, often merely by manpower and are speedily secured with simple fixing devices. Compare this with the difficulties experienced by the manufacturer of precast concrete components. Because of the weights involved, transport costs are high and handling in the factory and on site become expensive elements in the total cost of the components. This has led many concrete system builders to open temporary precasting factories on the sites of large projects, where transportation is eliminated and handling can be undertaken by the site tower cranes.

So much for timber's advantages; but what of its disadvantages? Resistance to the use of timber usually centres round two subjects; vulnerability to fire and its alleged lack of durability. In many people's minds these two aspects condemn wood as an impermanent material, good for very short-life cheap buildings, but for nothing else.

Nevertheless, timber in reality has remarkably reliable characteristics when exposed to fire, as will be illustrated in a later chapter. In spite of it being combustible, it will not collapse as readily as steel due to the effects of heat. The 1965 Building Regulations accept the use of timber, with satisfactory precautions, as a viable building material for sizable buildings of a permanent nature. The work of the Timber Research and Development Association has done much to dispel old misapprehensions about timber and fire, as it has also about timber and rot.

It is alleged that timber rots and is subject to infestation by wood boring insects; but rot only occurs where there has been poor detail design; or where the use of the material has shown a basic lack of understanding of timber's characteristics. In North America and Scandinavia, where there is a history of timber building, it has been rare in the past for wood to be treated with preservative. Here, appreciation of the material's characteristics has given it the accepted quality of permanence. With regard to infestation, it is usually sufficient to have periodic inspections and undertake remedial action when found necessary. In the UK there is an exception—that small area, the long-horn beetle country of north Surrey—where insecticide treatment is mandatory. Nowadays, with preservative treatment against rot and infestation available, there is no excuse for a lack of durability in timber. Nevertheless, the example of the many Middle Age timber buildings still in existence today should convince the doubtful that even in the days before chemical preservatives and insecticides, timber, correctly handled, could survive indefinitely.

If we examine the attitude of those countries with a tradition of timber building, such as Scandinavia, and North America; we find no prejudice against wood. In Canada, Building Societies, not the least nervous of bodies, happily give 95 per cent mortgages for private timber housing over a 25 to 35 year repayment period, and their experience of resale values are that such properties do not lose their values any more readily than those made of other materials. In short, in these countries the question of wood's durability is not fogged with emotional, muddled thinking. Even British Building Societies are becoming more amenable to the idea of timber building.

Wood has been around too long to be seen objectively. The emotion of the Great Fire of London still lurks in our national subconscious. Also wood lacks the glamour of a new material. There is a real danger it will be condemned to an oblivion it does not deserve.

New techniques and engineering advances have now revolutionised the use of wood. It is no longer used in great lumps, but has become a light, elegant, even a plastic material, capable of being moulded into gigantic lightweight shells, or coaxed into aspiring parabolic arches.

Much of what has been learnt in the past about timber was the result of trial and error over thousands of years—this learning has now to be unlearnt and re-learnt in terms, not of rules of thumb, but of scientific care and engineering skill.

Timber will burn, rot, twist and shrink, is variable in quality, limited in dimension, but all these problems one by one can be overcome by new techniques and an understanding of the material—and for all its faults it still remains one of the most exciting and 'good tempered' materials available. It is a material worthy of an objective assessment in terms of the twentieth century building scene. This is what this book is all about.

2 Timber in the past

Man's earliest form of habitation was provided by Nature. He sought out caves in which to hide from his enemies, keep hmiself warm and dry, protect himself from the rigours of the elements and set up his family unit. But caves were not numerous enough to satisfy the growing requirements of this ingenious animal. What was more, they were not necessarily situated where he wanted to live, namely near his food supplies and hunting grounds, and so Man was forced to construct habitations for himself and not rely on what was provided for him. His first building materials were the branches of trees, broken off by the wind—maybe at first laid against a cliff to form a one-sided wigwam, later laid against each other and tied at the top with supple twigs to form a free-standing tent-like structure, covered by turf, leaves or skins. No tools were needed for this type of building. The only cutting operation was that of trimming to length, which could be effected by burning (Fig. 2.1).

Fig.2.1 Early timber shelters.

This was the first time Man turned his hand to building—and the material he used was that which was most readily available. Another reason for early Man's use of wood was its ease of working. He tended to use the materials which were to hand. Where stone was to be found,

he laid stone on stone in their rough state; in the broad plains of the
Tigris and the Euphrates he used clay to make a crude form of sun-
baked brick; in Ancient Egypt, before the great building ages, he used
bundles of reeds bound together and placed vertically in the ground
and covered with clay; but wherever timber was available he used it,
because it had excellent structural characteristics and anything tougher
than wood was difficult to work with his primitive stone tools. But
these early tools were the very things which set Man apart from the
other animals. He used flint splinters for cutting, stone axes and even
flint awls for drilling rough holes. Gradually these were replaced by
bronze tools, and later by iron. A saw with hand-cut teeth made its
appearance and even a helical drill. It is interesting to note that the
tools, developed to work wood in pre-Christian times, are remarkably

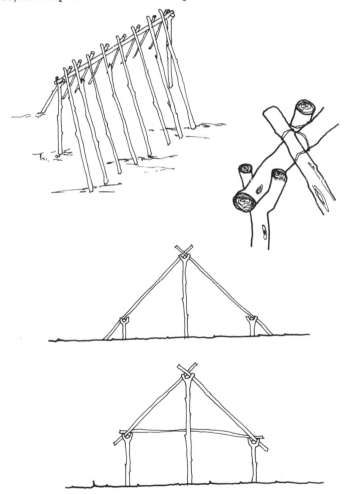

Fig.2.2 Early methods of building with timber.

similar to the present day carpenter's tools. The materials of which they are made have changed, but the principles of their application and design are the same.

Gradually the circular hut spread out to a rectangular building on plan with the branches sloping up to a ridge from either long side. Two posts were sunk into the ground at each end of the hut. Their tops had a 'Y' shape derived from the natural fork of a tree. These forks carried the ridge purlin which in turn carried the rafters with their lower ends stuck into the ground and their heads tied to the ridge purlin (Fig. 2.2). Skirting walls along the long sides developed in time and the first genuine cottage shape had been made—a remarkably universal form to be found in many primitive societies. As the spans of buildings increased, additional support was required for the rafters and this was supplied by further purlins, supported as in the English 'cruck' cottage, to be referred to later, by major timber frames. We shall see that the early European settlers in the New World many hundreds of years later were to use similar forms of construction.

In Switzerland, Italy and Ireland the remains of wooden lake dwellings have been found. These were no doubt simple structures similar to those described above, but this time set on wooden piles on the edges of lakes, much the same as are to be seen in primitive areas of Africa and the Pacific Islands today. They were sited in this manner to give additional protection to their occupants.

Early forms of timber joints must have been notches and roundings that allow one member to be housed into another and tied there. The hole and tenon must have appeared early, and when the first metal spiral drill had been invented, these joints would have been held together by a wooden pin driven through holes drilled through the joint (Fig. 2.3). But most of this is intelligent supposition, because there are few relics of pre-Christian timber buildings left to us. However, examples of primitive folk architecture in all ages seem to have displayed many remarkably universal characteristics which allow deductions to be made. Also, an indication of the way timber was used has come down to us in a curious manner—in imitations of timber structures in stone.

The most strange example of this is found in Ancient Egyptian architecture. Here, in a country where timber was scarce, but where granite, sandstone and limestone were abundant and brick was used usually faced with more durable stone or for strictly utilitarian structures, there have been discovered early rock-cut facades with obvious imitations in the hewn rock of a timber structure. They are strangely similar to rock-cut tomb facades in Lycia in Asia Minor, where half-notched frames and timber fixing wedges are all frankly imitated in the carving. This strange phenomenon has led some experts to suggest a common origin of these two peoples in some land where timber was plentiful.

Some authorities maintain that the Greek Doric style had its origins in timber construction. This has become a debating point over which

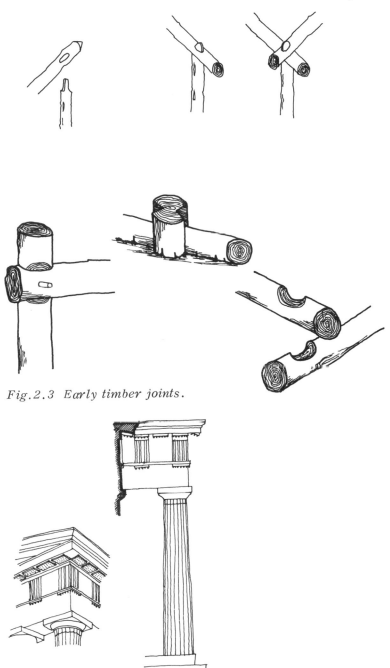

Fig.2.3 Early timber joints.

Fig.2.4 Doric column and entablature said to derive from building in wood.

tempers have become often frayed. Supporters of the timber theory claim that the Doric column and entablature derive from the wooden-built prodomus or porch of the Mycenaean palace. They cite the mutules (the inclined stone projections below the cornice) as stone reproductions of rafter ends, and the triglyphs over and between the columns as being imitations of beam ends (Fig. 2.4).

Pausanias in the second century A.D. described the temple of Hera at Olympia (B.C. 700) in these words. 'The style of the temple is Doric and pillars run all round it; in the back chamber (opisthodomus) one of the pillars is of oak'. This has led to the theory that the columns were originally timber and were later replaced by stone, although there are other theories that the wooden column had some ritualistic significance. However, the wider spacing of columns in the temple of Hera suggests an original timber superstructure. For many years the controversy over the alleged timber origin of the Doric style has raged backwards and forwards and there, no doubt, will never be a conclusion.

However, one fact is reasonably clear. The roofs of Greek buildings were constructed of wood, probably carrying stone slabs. Here is a fundamental point to remember about timber in construction. Although timber was usually relegated to domestic and less monumental architecture once Man learnt how to handle stone and brick, it still held its own as the long span material. While Man learnt the art of the arch and barrel vault in Roman times, he still usually used timber to construct the roof. And this situation existed right through the ages until iron and steel took over the long span work after the Industrial Revolution.

While the ecclesiastical authorities reared huge stone monuments to God in Gothic Britain, the vast majority of domestic architecture was carried out in timber. All, but the largest and most important residences up to the Renaissance, were basically timber buildings, from the rude 'cruck' cottage, to the imposing large half-timbered residences of the Elizabethan period.

The 'cruck' cottage is a development from the first primitive cottage form referred to earlier. It usually had a thatched roof with walls only at its gables. The 'cruck' construction was used extensively for barns and dwellings and sometimes incorporated a squat skirting wall on its two long sides (Fig. 2.5). As early as the fourteenth century there are definite signs of these 'cruck' cottages being subjected to a quaint form of prefabrication. After the crucks themselves were cut from the tree, they were made up in part where they were cut, then dismantled, transported to site and re-erected. It is interesting to note the holes in the feet of the crucks to aid in hoisting into position by crow bar and for temporarily tying together the feet of opposing crucks until the collar had been fixed.

The principle of half-timbered buildings, many hundreds of which are still standing today, was that of a very heavy post and lintel frame with an infill between the posts of brick or plaster. The quantities of

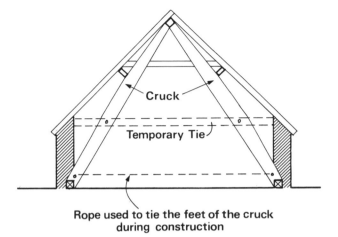

Cruck

Temporary Tie

Rope used to tie the feet of the cruck
during construction

Fig.2.5 'Cruck' construction with squat skirting wall.

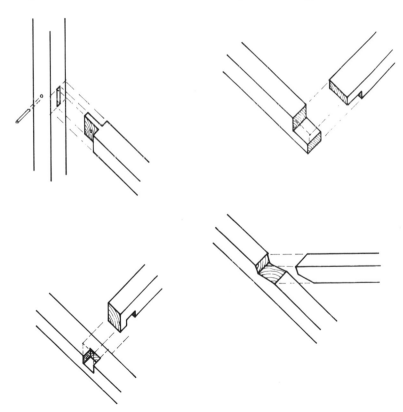

*Fig.2.6 Half timber building joints: the scarf, skew notch,
tenon, and cog.*

timber used were vast and the whole frame was held together by oak pegs.

Joints used were the skew notch, the scarf, the cog and the tenon held with a wooden pin (Fig. 2.6). These joints are still used today in very much the same manner.

At the same time a particularly English development was taking place in the roofs of great halls in the manor houses; this was the open timber roof. The Mediaeval carpenters realised the decorative value of timber roofs and a development commenced starting from the simple tie-beam roof, and leading through the trussed rafter roof and the arched forms of the collar braced roof, to the final intricasies of the hammer beam and double hammer beam roofs. Here was timber used both for its structural properties and for its innate beauty and versatility. These roofs produced some of the most exciting examples of timber's use in the whole history of construction. They demonstrate craft skills rising to great heights of vision and imagination.

Meanwhile, in Italy the first stirring of the Renaissance (which theoretically began about 1453) had been expressed by an interest in the great days of Greece and Rome and an outburst of original thought in a wide number of subjects.

Timber was still *the* long span building material and roof truss design was the main area of experimentation of these new free-thinking Italian engineers. The roof truss designed by Vasari (1511-1574) for the Uffizi Gallery at Florence is a typical example (Fig. 2.7). Here we see a king post truss employing iron straps at member junctions. The king post strangely did not connect with the tie beam directly. There was a 100 mm gap between the two members, the connection being made by an iron strap. Also Vasari was experimenting with splice joints to allow long tension members to be made up from several lengths of timber (Fig. 2.8). Obviously he was having difficulty in obtaining timber in the lengths he required for his designs.

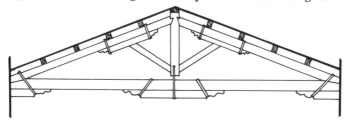

Fig.2.7 *Roof truss designed by Vasari for the Uffizi Gallery, Florence.*

Andrea Palladio (1518-1580) made the next advance in timber truss design when he applied the principle of the rigid triangle truss to bridge structures. His first timber bridge was a stringer bridge near Bassano over the River Brenta (Fig. 2.9), but when the spans involved were too great for a stringer bridge he started experimenting with

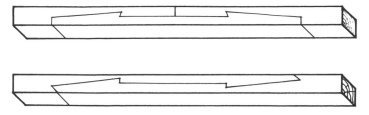

Fig.2.8 Vasari's splice joints.

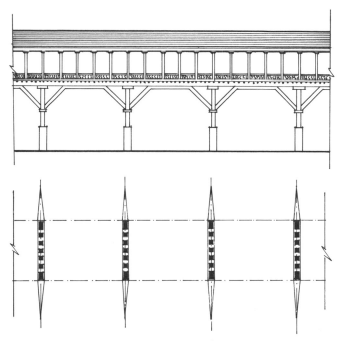

Fig.2.9 Palladio's stringer bridge over the River Brenta, near Bassano.

trusses. His first truss bridge (Fig. 2. 10) was over the Cismone River and employed metal straps. He later went on to apply similar principles to bridges of wider span and curved form.

At the same time, in France, Philibert de l'Orme (1510-1570) was experimenting with building up rafters out of individual planks. Not only does this method relieve the designer of some of timber's dimensional restraints, but it also allows for the constructing of curved members. The important example of the application of this technique is the apse of the Church of the Annonciades at Antwerp (Fig. 2. 11). It is interesting to note that the dimensional limitations of available timbers was one of the main difficulties designers were endeavouring to overcome. This situation is even more marked today.

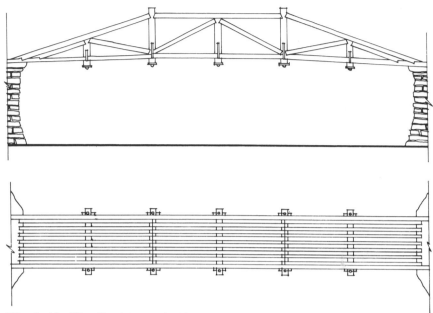

Fig.2.10   The first truss bridge by Palladio using metal straps.

Advances in combining timbers to produce long span structures were made by Faustus Verantius from Dalmatia who was at one time a bishop and secretary to the Emperor of Austria. Towards the end of the sixteenth century he was producing timber bridge designs using relatively small timbers combined with bolts and tie plates. In later designs he even introduced iron tie rods.

Palladio, de l'Orme and Verantius, working on their respective inventions, laid the foundation for large framed structures and modern truss designs.

The Renaissance was late coming to Britain. Gradually classical details began to creep into Elizabethan and Jacobean architecture, but it was with Inigo Jones (1573-1652) who had studied in Palladio's native town of Vicenza, that the full effect of the classical rebirth was evident.

In 1683 an English edition of Palladio's 'First Book of Architecture' was published. This edition contained an appendix by the English architect, William Pope, dealing with the framing of 'all manner of roofs'. This appendix added nothing significant to the rigid triangular truss used by Palladio.

In Britain for many years there was no major advance in the use of timber. The next development was to come from the other side of the Atlantic. It is, however, noteworthy that timber was used for major bridge building as recently as the 1850's in England.

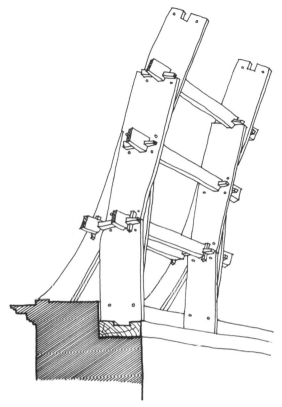

*Fig.2.11 The apse of the Church of the Annunication at
Antwerp shows the application of de l'Orme's technique for
building rafters out of individual planks.*

I. K. Brunel constructed viaducts for the Cornwall and West Country
Railway Companies in timber from standardised designs. There was
the 'fan' system, where struts radiated fan-wise from masonry piers
(Fig. 2. 12), and there was the light timber truss method spanning between piers. The latter design allowed for the removal of any member
and its replacement with a minimum of difficulty. Several of these
viaducts survived until into the twentieth century.

It is one of those strange accidents of history that has led to the direct
line from mediaeval timber framed houses and open trussed roofs to
present day timber engineering being traceable in the architecture, not
of Europe, but of the New World.

The early settlers took across the Atlantic ideas of cottage building
they had known at home. Some of the very early settlers' houses in
New England were made by driving stakes into the ground and then
interweaving them with branches and covering the whole thing with a
clay daub. The cruck method of construction made its appearance on

Fig.2.12 Old Liskeard Viaduct: timber struts radiating fan-wise from masonry piers. Copyright: Western Region, British Rail

that side of the Atlantic, as did also the Swedish type of log house which first made its American debut in Delaware. This latter was not a frame construction, but a straight-forward load bearing system in which the wall was made by laying tree trunk on tree trunk with halved overlap joints at the corners of the building.

By 1630 houses of the half-timbered type of frame construction began to make their appearance. It is even recorded that in 1624 the English took to America a 'panelized' house of wood which was dismantled, moved and re-erected. This might almost be considered to be the first of many exported prefabricated structures to leave Britain.

The New England house was usually constructed of a hand-hewn oak frame. At first joints used were simple butt joints with wooden pins, or the ends of the beams were bevelled and fitted into notches in the posts. Later the mortise-and-tenon joint with wooden pins became almost universal. The posts usually were at about 600 mm centres and ran the full height of the building from a timber cill on a masonry foundation. At first horizontal members were used on their own, later diagonal bracing members were added. Infill between the posts started off as wattle and daub and graduated to brick. All this is a direct reflection of the earlier British developments. The American contribution though was the clapboard cladding which made its appearance in the

18th Century. Sheathing was used behind the clapboard as a nailing base and the infill disappeared.

Through the nineteenth century in America, timber was associated with the vernacular tradition, not with architectural distinction. Monumental work tended to be carried out in masonry. Timber became associated with utility. Nevertheless, there was a continuous tradition of timber building in America and it was probably because of this that the next big step forward in timber construction was taken in American and not in Europe, where the tradition generally had been interrupted. The American innovation was the 'balloon frame' (Fig. 2.13). The term 'balloon frame' was originally derisive, because of the lightness of this new method of building. It was also known as 'Chicago Construction'.

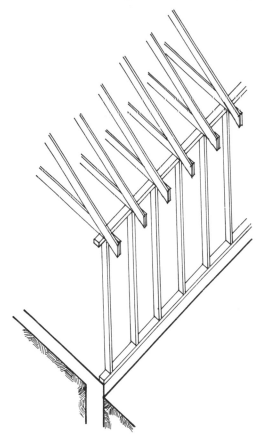

Fig. 2.13 The 'balloon frame', an American inovation.

Augustine Deodat Taylor, architect and builder of Hartford, Connecticut arrived in Chicago in June 1833 where he was commissioned to build St Mary's Church. Construction started in July 1833 and was completed

in October 1833. This was the first building in the new framing system
derived by Taylor from the New England frame. However, instead of
the substantial posts set wide apart of the original, Taylor used a large
number of closely centred studs and joists with horizontal cills and
plates framed together at the edges of the floors and roofs. Most of
the timber posts used were only 100 mm × 50 mm and spacing normally was at 400 mm. Only three or four sizes of timber were used in
the whole construction.

An interesting parallel development was the patented truss of Ithiel
Town in 1820 (Fig. 2.14). This timber truss was made up of closely
spaced intersecting diagonals forming the web with horizontal members
of two or more parallel timbers on either side of the web members.
These trusses were used for bridge construction, one on each side of
the roadway; and were often roofed. Timbers used were mostly of
domestic size and the connections were nailed as in the balloon frame.

Fig.2.14   The patented truss of Ithiel Town

In was the introduction of mass produced nails that made the balloon
frame practicable. Quickly this new technique was adopted for residential buildings throughout North Amercia. It proved simple, strong
and light and had the advantage of spreading all floor and roof loads
equally over many support members. Two models of the balloon frame
were exhibited at the Paris Exposition of 1867. Architects, however,
tended to ignore the development, probably because of the feeling that
timber was a second class structural material: but be that as it may,
it opened the way to the whole of timber building developments up to
the present day. The significance of Augustine Deodat Taylor's contribution cannot be over-stated.

One of the side effects of the balloon frame was the launching of the
timber prefabricated building industry in North America—maybe in

a small and scattered way at first, but nevertheless timber prefabrication had suddenly become viable.

By 1860 several firms in Boston, New York and Chicago were prepared to ship prefabricated sections of framing to any rail terminal in America. Sections of building were joined together by nails or bolts. It was claimed that the erection was a do-it-yourself operation.

The ready-built structure manufacturing business by 1880 had become a major industry in North America. Houses, farms, warehouses, bathing houses, even railroad stations were made in component form in factories and shipped to all parts of the country. In 1897 a New York company was shipping prefabricated timber houses to Alaska for the Klondike Gold Rush.

During the early twentieth century in America there was a falling off of interest in prefabricated building, due in part to architectural opposition and in part to restrictive practices of the building trades unions. The tradition of timber domestic building, however, continued uninterrupted.

In Britain the balloon frame had an effect, albeit a relatively minor one. In 1840 Britain was in the frenzy of the Industrial Revolution, absorbed by the exciting innovations of the engineers and cutting down the forests at a terrifying rate in order to feed the hungry revolution. Generally timber was used structurally only for the floors of domestic buildings and the roofs of domestic as well as more prestige buildings. However, there was a fining down of timber sizes, maybe resulting from the balloon frame.

Peter Thompson in 1840, influenced by the Chicago experiment, started manufacturing prefabricated timber churches. His first completed structure was the Church of St John the Baptist, Kentish Town. In 1845 he wanted to erect more churches in London and, even in spite of difficulties with the Metropolitan Building Office, he succeeded in erecting six further churches.

Europe at large gave little productive thought to timber as a building material at this stage. Only in Scandinavia where, like the United States, there was a tradition of timber building, is seen experimentation in shop fabrication of building sections. In Sweden by 1930 there were a considerable number of timber prefabricated housing methods in existence, some of which were even exported to Britain.

Although there was considerable interest in factory methods of building in Britain during the nineteenth century, and a considerable number of factory-made buildings were produced, such as those exported to the 1849 Californian Gold Rush, corrugated iron churches for many places in America, a hospital (designed by I. K. Brunel) for the Crimea War, they were all undertaken in metal.

Industrialisation had started to come to the building process, but it was going to be some time before the obvious material for this context—wood—was going to be given its opportunity.

Man's oldest building material was still suspect—or if not suspect, ignored. In fact, timber was waiting for associated developments to occur, before it could take its next step forward.

As we have seen, timber (with a few notable exceptions such as Brunel's viaducts) had been relegated to domestic purposes as soon as Man had acquired the skills to use other, more difficult to handle, materials; first stone and brick, later iron and steel. To handle timber with the engineering skill demanded of a material as sophisticated as it is, its characteristics had to be explored, its strengths and its weaknesses probed; methods had to be discovered of overcoming its dimensional limitations. The same degree of development effort needed to be put into the exploration of timber as was needed to move from cast iron to high performance steels. Iron progressed quickly because its uses were multifarious and development effort was directed from many sides, driven on by the Industrial Revolution. A similar effort was needed in the case of wood if it was to become a material of the latter half of the twentieth century.

3 Development in timber engineering

Carpentry and timber engineering have only one thing in common—the material they both use, wood. Otherwise, the two are completely different in philosophy.

The carpenter, as an exponent of one of the oldest craft skills, handles timber in an empirical manner, basing his treatment on years of trial and error and a sympathy for the material which generations of experience have produced. The craftsman knows instinctively how large a timber component should be to perform a particular task—the size of a joist to span an opening, the girth of a post of a certain height to support a certain load. This inherited experience led to the compiling of rules of thumb to guide succeeding generations. Such rules of thumb are those to determine the size of domestic floor and roof joists and which are expressed as:

$$\text{depth of floor joist in inches} = \frac{\text{span in feet} + 2}{2}$$

and

$$\text{depth of roof joist in inches} = \frac{\text{span in feet}}{2}$$

These guide rules and similar were discovered to be safely adequate, little or no account being taken of the quality of the timber used. Safety in these terms was therefore often synonomous with wastefulness.

In many respects the traditional carpenter can be compared with the gothic stone mason, who experimented daringly with stone to produce ever more light and aspiring structures, learning by his or other's failures, discovering about the material as he went along.

The advantage that the stone mason had was that the material with which he was dealing was in many respects more consistent, type for type, than that of the carpenter. So the refining process of the timber structure, from the tree trunk to the light timber frame, took longer, being dependent, as it was, on related skills such as the production of mass produced nails, high strength glues and more sophisticated mechanical fixing devices.

The timber engineer, in contrast with the craftsman, applies engineering principles primarily and intuition secondarily. To do this he needs quantifiable facts on which to base his activity. The application of engineering principles to timber was slow, partly because of a lack of understanding of the real strength characteristics of wood and how to recognise reliable from less reliable wood, partly because of the feeling that timber was a second class material from a structural standpoint.

Timber had been around so long that there was a tendency to ignore its capabilities. It had been used through the years as a material able to span wide openings, before reinforced concrete had been thought of, before cast iron and later, steel, had come to oust it from many of its traditional strongholds. The newer materials tended to attract the engineer's attention and the traditional materials—stone, brick, and timber—got left behind. But of these traditional materials, one in particular, timber, had got an engineering potential that the others have not and it was the recognition of this engineering potential that has taken so long.

But, as we have seen in the previous chapter, in the same way the 'balloon frame' had to await the introduction of the mass produced nail, so timber had to await two developments before it could be treated as a fitting material for use with engineering skill. These developments were:

(a)   stress grading
(b)   synthetic resin adhesives

But before considering these two developments, it would be advisable to re-state the characteristics that make timber a worthy material for engineering treatment. These characteristics are:

(1)   Good strength to weight ratio which makes it ideal for factory-made structural components, allowing large strong assemblies to be easily handled and transported.
(2)   Ease of working with simple tools and machines
(3)   Assembly in simple jigs
(4)   Easy fixings with nails, screws, bolts and adhesives.

In short, wood is strong, adaptable and easy to handle. It has, however, severe dimensional limitations. From its natural origins it converts easily into sizes, which are restricted both in cross sectional area and length. Although BS 4471: 1969 'Dimensions for softwood' give basic size tables which show softwood cross sectional dimensions varying from 75 mm × 16 mm to 300 mm × 300 mm and lengths between 1.800 m and 6.300 m in 300 mm increments, only a limited number of these sizes are readily obtainable in quantity and at a reasonable price. The most common sizes from which most sections are derived are 100 mm × 50 mm, 150 mm × 50 mm and 200 mm × 75 mm, while lengths in excess of 2.400 m or 2.700 m are relatively rare.

In addition, timber in its natural form makes a poor 'sheet' material. With a maximum width of 300 mm difficult to obtain, and a minimum thickness which is viable at that width of 25 mm, wood is highly restricted. Even then this long thin cross section is likely to warp in practice. It was essential, therefore, to find a method of converting wood into a viable sheet material; plywood and particle boards have resulted. The cross sectional limitations needed to be tackled as well. Obviously, these could only be overcome after the development of high stress glues; and the laminated beam resulted. Nevertheless, the full strength of glues were only going to be fully exploited if the actual strengths of the pieces of wood to be joined together were understood. This, in fact, is the root problem of modern timber engineering and why stress grading is so important.

STRESS GRADING

Stress grading will be discussed in detail in Chapter 7. Suffice it to say here that before wood could be treated seriously as a material capable of being handled with precision and economy (in fact given a full engineering treatment) it was necessary to establish its structural reliability.

Timber is a variable material, conditioned by the vagaries of its growth. The succession of frosts and winds, late or early springs, wet or dry summers, all leave their mark on the growing wood and its ultimate strength.

Selection of timber is at present mostly undertaken visually, applying standards laid down in CP 112 where strength grades are determined by numbers and sizes of knots in particular sized timbers; incidence of gum pockets, splits, distortion and slope of grain. This appearance grading takes into account only the more obvious evidence of strength or weakness, and so the standards must be compiled accordingly, erring always on the side of safety. So often the process degenerates into finding the largest knot in a piece of wood and estimating whether it is smaller or larger than a simple fraction of the surface areas on which it occurs.

Recently mechanical stress grading has made advances, particularly in North America, Australia, South Africa, and the United Kingdom. This is based on a relationship which exists between the strength of timber and its stiffness, or the amount it deflects under load. A machine has been devised to impose a small load on timber and measure its deflection: the larger the deflection, the weaker the timber.

The machine grades irrespective of the species of the timber, whereas visual grading depends on knowing what is the species of timber that is being graded. This can sometimes cause problems and can lead to mistakes.

The effect of machine grading will be to establish much greater definition of grades and a consequent fining down of engineering safety factors and reduction in wasteful over-design.

SYNTHETIC RESIN ADHESIVES

Without reliable adhesives of great strength, the whole present day plywood manufacturing industry would not have been possible; laminated timber structures would never have been developed; modern jointing techniques that allow the use of ever smaller pieces of wood would have been impractical. Maybe it would not be an exaggeration to say that the most significant single development that has forced wood out of the carpentry straight jacket has been the production of synthetic resin adhesives and the whole complicated technology of glue manufacture and use.

Synthetic resin glues were first used in the Second World War, when urea formaldehyde was chosen for its resistance to water for use in the aircraft industry. In some applications a gap filling characteristic was required because of the difficulty of applying the necessary pressure during the setting of the glue in some positions in an aircraft. Modified urea formaldehyde glues with a special hardening agent solved the problem. The Mosquito aircraft was the most well publicised wooden aircraft whose existence would not have been possible without synthetic resin adhesives.

What is synthetic resin? It is a resinous substance prepared by a chemical reaction from simple compounds. No synthetic resin, so far as is known at the moment, has a counterpart in nature.

Glues set in three ways:

(1) Loss of solvent. This process is reversed by merely adding solvent or water. Such glues are not permanent by nature, and where the solvent employed is water, are adversely affected by humid conditions. Some glues that set due to a loss of solvent are sodium silicate, starch and cellulose etc.

(2) Cooling from a molten state. This again is a reversible process. Bitumen adhesives are a good example. Gelatine adhesives set partly by cooling and partly by the loss of the solvent—water.

(3) Permanent chemical change. Synthetic resin adhesives belong to this group. The processes are irreversible and are mostly the combination to a greater or lesser extent to two methods—the addition of a hardener which completes a chemical reaction or the application of heat. The example of the first method (which can be speeded by the second) is urea formaldehyde (UF); of the second method, melamine formaldehyde (MF) and phenol formaldehyde (PF).

To the extent that urea formaldehyde, without fortification by MF or resorcinol additives, is somewhat less resistant to the effects of hot water, its setting process could be considered slightly reversible, but nevertheless perfectly adequate for normal building uses.

The principle of the setting of synthetic resin adhesives is that the chemical process which would result in the solid resin and which is started during manufacture is halted at an intermediate stage. The completion of the process takes place during the glueing operation and the chemical reaction is re-started by the addition of a hardener or the application of heat. Generally synthetic resin glues are expensive. Broadly the more resistent to water, the more expensive; but cost is relative to the physical properties required of the joint. Extenders can be included to lower the cost of the glue line. Such extenders are starch, dried blood, wood flour and china clay. Starch is particularly good as an extender for hot pressure adhesion. Semi gap-filling extenders are wood flour and china clay. These take up irregularities to a certain extent in the adherents, but the addition must be kept down to between 5 per cent to 10 per cent, otherwise the capabilities of the glue line are affected.

The most normally used synthetic resin glues in timber engineering are:-

Urea formaldehyde (UF)

This is the most widely used glue in the wood industries. It is made by heating one molecular portion of urea with about two of formaldehyde under weakly acid conditions. It is supplied as a powder which, if kept in cool conditions, has a good shelf life. A hardener is applied at the glueing stage and the chemical reaction is recommenced and completed. Setting time depends on the acidity of the hardener and sometimes a retarder has to be used to lengthen the glue's pot life.

The glue line is insoluble in common organic solvents and resistant to acids and alkalis. It is water resistant up to temperatures of 80°C, after which its resistance falls off badly and at 100°C (if not fortified) it has no strength at all. UF does not have a high resistance to heat. It can be fortified with melamine or resorcinol formaldehyde.

Melamine formaldehyde (MF)

This adhesive can be cured by heat without the addition of acid hardeners. It does not make a satisfactory cold setting adhesive. It should be used above 60°C, mostly for hot press glueing. MF is completely resistant to boiling water. It has superior chemical and physical properties to UF, but is no stronger.

Phenol formaldehyde (PF)

This is another hot setting adhesive. It can be used as a liquid glue or in a dry resin film form, using paper as a carrier. PF needs temperatures in excess of 100°C for curing, usually with the addition of an alkali hardener. Acid hardeners can be used for cold setting, but it is a practice that does not prove too satisfactory. PF is resistant to boiling water and is used in the production of exterior quality Canadian fir plywoods.

Resorcinol formaldehyde (RF)

Formaldehyde itself is used in this case in place of a hardener. By its use the resin can be converted without heat into a completely infusible material which is resistant to boiling water.

There are other synthetic resin adhesives in frequent use, but the four detailed above are those most often encountered in timber construction. Other synthetic resin adhesives include PVC adhesives, used mostly for adhering surface coatings, polyvinyl aldehyde adhesives, used mainly for adhering glass and metal, and ethoxyline resin adhesive (Araldite) which is a curious glue, whose chemistry is still not fully understood. Araldite will stick almost anything to anything, either as a hot or cold setting process—the hot process producing a joint of greater strength. Araldite is resistant to boiling water, acids, alkalis and many organic solvents. It is, in fact, the nearest approach so far to a universal glue. One word of warning is necessary. In all these glues the control of moisture in the timber is essential. Therefore these are mainly glues to be used under factory conditions and not on the site. Even in the factory the moisture content should be controlled within the range 12-15 per cent. RF can be used up to 26 per cent moisture content with only a small falling off of strength. At that moisture, UF is drastically affected.

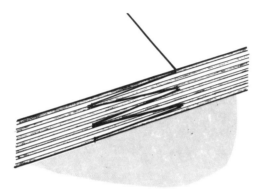

Fig. 3.1 Finger joint.

With very strong adhesives available the way was open to look into the traditional methods of jointing timber and appraise these in present day terms. The finger joint, among others, has resulted (Fig. 3. 1). This is a joint of large surface area compared with the cross sectional area of the wood jointed. It is used for end jointing short pieces of wood to form longer components. Such joints are cut with great accuracy on special machines and the effect of the joint is a strength of at least 80 per cent that of the clear timber. At present the Princes Risborough Laboratories of the Building Research Establishment are investigating a die formed end jointing machine to press miniature finger joint profiles into softwood sections.

PLYWOOD AND ITS USES

The production of synthetic resin adhesives gave a great impetus to plywood manufacture. Previously plywood had been severely limited in its application and very vunerable to moisture. As many of the present day developments in timber engineering depend on the use of durable plywood, it can be understood how important thoroughly reliable adhesives are.

Plywood is made up of thin veneers of wood bonded together with adjacent layers having opposing grain. The veneers are always odd in number and vary in quantity with the thickness of the completed sheet. BS 1455: 1963 'Plywood manufactured from tropical hardwoods' requires all plywoods greater than 10 mm thick to be made with not less

than five plies. The opposing grains of veneers give plywood great
strength; in fact it is a hundred times as stiff as a steel plate of equal
weight. Additional characteristics are:

> Movement is minimised (the more plies, the less movement): it
> will not split: it has no line of cleaverage and therefore will with-
> stand knocks: it can be bent: it can be faced with metal or lamina-
> ted plastics: it has the thermal conductivity of wood (average
> 0.138 W/m deg C): it has a high nailing strength.

Principal species of timber used for plywood manufacturing are Baboen,
Basswood, Beech, Birch, Douglas Fir, Gaboon, Ilomba, Lauan, Mahogany,
Meranti, Parana Pine, Poplar, Sapele, Serayah, Utile, Spruce, and Virola.
For structural purposes the most generally used woods are Douglas
Fir coming from Canada and Birch from Finland.

The manufacturing process of plywood is as follows. Selected logs
about 2550 mm long called 'peelers' are stripped of their bark. They
are then set up in a giant lathe where a steel blade peels the veneer
off the log. A Douglas Fir 'peeler' can yield as much as one mile of
veneer (Fig. 3.2).

The veneers are then pre-shrunk in driers and checked for moisture
content. Then laying up of the plies takes place with the face ply
chosen as appropriate to the grade of plywood being produced. The
plywood is then bonded in a press. In the case of PMBC Exterior
quality Douglas Fir plywood the temperature of the press is 148°C,
the pressure applied is 1.4 N/mm² and the glue is phenol formaldehyde.

Glues used in plywood manufacture are selected for the degree of
exposure that the plywood is going to have to withstand. Standards are
laid down in BS 1455: 1963. All exterior quality plywoods, or type
WBP (weather and boil proof) are bonded with a synthetic resin adhesive other than UF. Interior plywoods, or type INT, can be bonded with
soya bean, casein or blood albumen glues. Between these two extremes, there are two other types listed in BS 1455: 1963; type MR
(moisture resistant) and type BR (boil resistant). These types are
suitable under most circumstances, but not when exposed to extreme
weather conditions: the adhesives are selected appropriately.

Plywood is also graded in accordance with the standard of the face
veneers. At the lower end of the scale is sheathing grade, produced
for roof and wall sheathing and with limited open defects permissible
on one or both faces. At the upper end of the scale is 'good two sides'
grade with highest grade veneer to both faces. Below this is 'good/
solid' with highest grade veneer one side and relatively good veneer
to the other; 'solid two sides' with relatively good veneer to both sides;
and 'solid one side' with good veneer on one side only. In addition,
there is 'marine grade' which is a high grade resistant plywood with
all its interior veneers solid.

Plywood is supplied in lengths of 1.800 m, 2.400 m, 2.700 m and
3.000 m; widths of 600 mm, 900 mm and 1.200 m and thicknesses from

Fig. 3.2 Veneers are peeled by huge lathes. (Copyright: Council of Forest Industries of British Columbia).

2.7 mm to 50 mm. All these sizes are not necessarily readily available. The first quoted dimension of a sheet is that parallel to the grain of the outer veneer. Long and short grained sheets are available.

It is possible to obtain sheets of plywood to greater dimensions up to 3.600 m × 12.000 m. These are produced by scarfe or finger jointing standard sized sheets with a joint gradient of 1 : 10 of the thickness. Such joints can increase the flexural stiffness of the sheet, rather than weakening it.

The uses of plywood are infinite; illustrating its importance in the building industry at large and to modern timber engineering in par-

Fig.3.3 Stressed skin panels (Copyright: Council of Forest Industries of British Columbia)

ticular. In addition to its use in formwork for concrete construction, it is used extensively for sheathing and cladding, where advantage is taken of its great racking strength. Plywood is vital to developments which are discussed in greater detail later in this chapter; box beams, rigid frame structures and shell roofs.

One of the most important uses of strong and reliable plywood is the stressed skin assembly. By combining timber framing with a plywood sheet and taking advantage in the calculations of the strength of the plywood and the integral action of framing and plywood, there is a significant reduction in the sizes of the timber framework and large spans can be tackled with light assemblies (Fig. 3.3). The only proviso is that the jointing between the panel and the framework must be strictly controlled and one hundred per cent reliable. Usually (but not absolutely invariably) these joints are glued and nailed and really should be undertaken in factory controlled conditions.

Stressed skin panel technique can be applied to box beams, floor and roof deck units, bracing wall panels, in fact to most high performance frames in a modern timber building. Its most spectacular use is probably in the folded plate roof where pitched stressed skin panels are rigidly jointed together to produce an exciting and economic long span roof construction (Fig. 3.4).

Space frames, consisting of rows of hollow plywood tetrahedrons covering spans of up to 30.000 m square, provide economic two-way span

Fig. 3.4 Folded plate roof using pitched stressed skin panels. (Copyright: Council of Forest Industries of British Columbia).

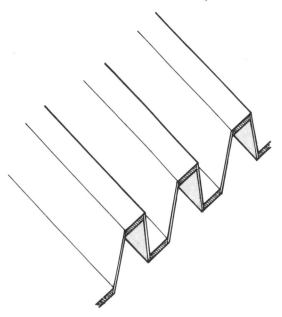

Fig.3.5 Trofdek roof of trough configuration using plywood and softwood chords.

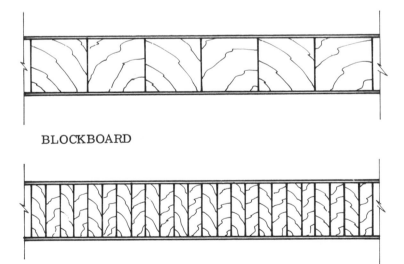

BLOCKBOARD

LAMINBOARD

Fig.3.6 Laminboard with a core of veneers sandwiched between facing veneers.

solutions to long span roofing problems, easily and speedily erected by manpower alone.

Another use of plywood in spanning wide spaces is the Trofdek roof, which combines the strength of plywood and softwood chords in a factory made unit of a trough configuration (Fig. 3. 5).

There are other types of board produced by lamination techniques. They are known as blockboards and laminboard. Both are expensive, but combine strength with great stability and a resistance to warping. Blockboard has a core of wood strips (19 mm to 28 mm wide) glued together and sandwiched between facing veneers. Laminboard is a more high grade blockboard with a core made up to 1. 5 mm to 7 mm wide veneers sandwiched between facing veneers (Fig. 3. 6).

PARTICLE BOARDS

An acceptable sheet material, sometimes making use of wood, or other vegetable media is covered by the generic term 'particle boards'. They are all boards made from wood chips, or flax, or hemp shrives bonded with synthetic resin or an organic binder, and are referred to as wood chipboard, flaxboard or hempboard.

The significance of this development is the use of wood chippings. These can either be the waste products from other sections of the timber industry, or the result of thinings from afforestation projects. As timber is a valuable material and likely to become more so in the next few years, the ability to achieve minimum waste is essential. Chipboard is therefore a very important development and one which will no doubt be the subject of major advance as time goes on.

Sheet sizes of particle board, according to BS 4606: 1970 'Co-ordinating sizes for rigid flat sheet materials' are in length 1. 800 m, 2. 400 m, 2. 700 m and 3. 000 m, and in width 600 mm, 900 mm and 1. 200 m. Thicknesses are 10 mm, 12 mm, 15 mm, 18 mm, 22 mm, and 25 mm, but other thicknesses varying from 4 to 40 mm are obtainable.

The disadvantage of particle boards generally is their sensitivity to moisture. Chipboard, for instance, must be kept at a moisture content below 14 per cent if unacceptable shrinkage on drying is to be avoided. At a moisture content of 20 per cent physical deterioration of the board can occur. Current research is concentrated upon improving chipboard's resistance to water with improved binders and edge protection and in-built fungicides and water repellants. If this proves successful, chipboard would be able to be used in external situations or those internal situations where condensation might be a danger. At present chipboard is limited in its structural application because of these difficulties, but high density chipboard ($680-750$ kg/m^3) is already used extensively in place of traditional tongued and grooved floor boards, while medium density chipboard ($550-650$ kg/m^3) is used for roofing, partitions and fitments.

Chipboard is manufactured as single, three or multi-layer boards. Single layer boards have a consistent density throughout their thickness. Three layer boards have high density outer layers. Multi-layer boards have a high density core in addition to the outer layers. Three and multi-layer boards give a better strength/weight ratio and stability than single layer boards. Single layer or multi-layer boards have better performance where edge fixing is necessary.

EXAMPLES OF MODERN TIMBER TECHNOLOGY

The whole of modern timber engineering techniques are based on the combination of the following:

(a)   timber of a predictable performance,
(b)   glues of a calculable strength,
(c)   high grade plywood,
(d)   mechanical fixing devices: nails of infinite variety and strength including twist nails; screws; coach bolts; bolts; and a wide range of proprietary connectors such as the Gangnail (a multi-toothed connector plate applied mechanically by a special machine) (Figures 9/32 and 9/33), the pre-drilled connector plate for nailing in the traditional manner, toothed connector plates used in conjunction with bolts, split ring or shear plate timber connectors (Fig. 3. 7).

Developments range from highly economic standard pitched roof trusses produced with mass production line methods to one-off laminated parabolic arches.

We shall examine just three of the major areas of development.

Beam design:

Problem. To span wide spaces using a minimum of timber, thereby placing the timber where it does the greatest work, namely in the upper and lower edges of the beam. The traditional solution is the solid beam, where the majority of the sectional area of wood is doing practically no work at all. In fact a large area of the wood, by its own dead weight, is hindering rather than aiding the solution.

Solution. *Timber lattice beam* (Fig. 3. 8). This is made up of continuous top and bottom members (chords) held apart by diagonal or vertical members. The chords can be made up of more than one piece of timber employing end to end jointing with glued finger or scarf joints. The stiffness of the beam (the degree to which it will deflect under load) is proportionate to its depth. Points of high stress are in the joints between the braces and the chords as the area of the joint, and consequent glue line, is restricted. As a result, steel strip reinforcement (either plating the joint or inserted into it) is sometimes used to overcome this critical situation which otherwise would have to be tackled by increasing the timber sizes with resulting wastefulness. Spans up to 13. 500 m can be achieved with depths of beams up to 900 mm.

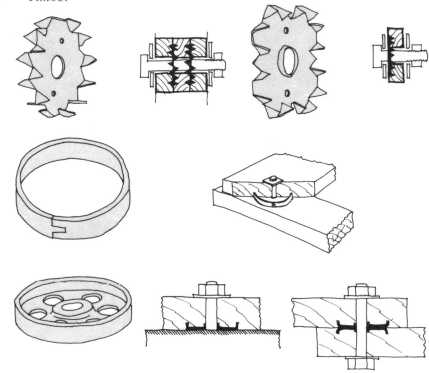

Fig. 3. 7   *Toothed, split ring, and shear plate timber connectors.*

Plywood box beam (Fig. 3. 8). Like the lattice beam the chords are held apart by vertical soldiers, but set at greater spacing. The web action is achieved by plywood sheets on both sides of the chords. These are glued and nailed to produce a stressed skin effect. The timber content in box beams is less than in lattice beams and the jigs in which they are made, are simpler, and no steel reinforcement is necessary. The achievable spans are similar to lattice beams with similar depths.

Corrply beam. (Fig. 3. 8). This is a proprietary type of beam manufactured by Rainham Timber Engineering Co Ltd. It illustrates admirably the skilful use of material to get the best performance out of as little wood as possible. The chords are finger jointed into continuous lengths which are fed into a machine that grooves them in a sine curve. Plywood on a 100 m spool is fed through a section of the machine where its edges are profiled and spread with resorcinol formaldehyde glue; the machine then inserts the plywood into the curved grooves in the upper and lower chords and the beam is complete. The process is entirely automated and the beams are made in one continuous length and are cut to size on leaving the machine.

The Corrply beam has great rigidity and lateral stability. The curved form of the plywood gives stiffness and prevents buckling of the web.

Spans achievable are from 3.000 m to 13.500 m with depths from 300 mm to 600 mm.

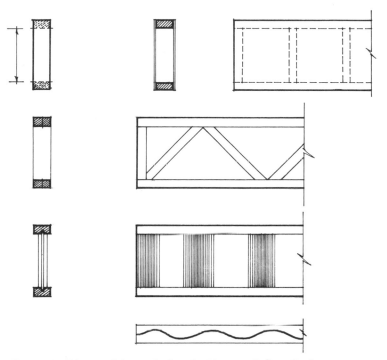

Fig.3.8 Plywood box, timber lattice, and Corply beams.

Rigid frame construction

Problem. To construct wide clear span buildings with a small erection crew and simple tools.

Solution. The origin of rigid frame construction was in 1954 in Canada. The idea was developed by the Plywood Manufacturers of British Columbia and was tested by the University of British Columbia Agricultural Engineering Department. The principle of this form is the construction of a series of three-hinged arches consisting of four timber members joined at the haunch and crown with plywood gussets (Fig. 3.9). All fastenings are undertaken by nails only. Frames are spaced at 600 mm, 800 mm, or 1.200 m centres (the first two being the most economic) and the whole building is sheathed with plywood to form an extremely rigid and strong structure, the plywood bracing the whole building.

The construction results in buildings of wide clear spans without the use of braces or trusses. Simple, standardised frames are constructed in the factory or on site. Because of their lightness, they are easy to erect by manpower and without the aid of plant. The result is a fast

enclosure of space (a two man team can enclose 50 m² per hour),
economy (only 50 per cent of the amount of labour required for an
equivalent traditional building) and versatility.

In spans up to 12. 000 m, solid timber members are suggested, above
12. 000 m, top and bottom frames are employed with plywood or lattice
webs. Laminated beams also can be used.

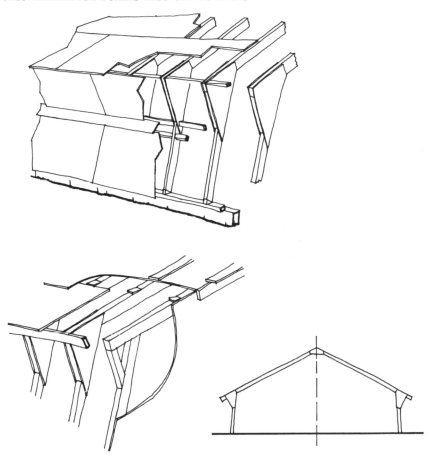

Fig. 3.9 Detail of rigid frame construction.

A development called the HB system could be mentioned at this point.
It was invented by a Swedish professor, Hilding Brosenius. It is a
specially built up construction using only nail connections. It consists of laminated top and bottom flange members divided by a web
made up of two layers of boards, at right angles to each other and both
at 45° to the flanges. This construction produces a very high shear
resistance and avoids the use of plywood.

Fig. 3.10 Norton Swimming Pool, Norton, Yorkshire, showing Glulam members. Copyright: Rainham Timber Engineering Co. Ltd.

Glued laminated timber structures

Problem. To overcome the dimensional limitations of timber in its more usual form and the limitations of shape of the finished component.

Solution. The process of glueing together several small sections of timber to produce a component of greater cross sectional area than is normally obtainable (Fig. 3. 10).

In 1905 a German, Otto Hetzer, took out a patent for what became known as the Hetzer System, which involved the glueing together of layers of timber with casein glue to form structural components. There was early interest in the process in Switzerland (cf. the tower of the University of Zurich, 1913), but it was the development of synthetic resin glues in the Second World War that gave the boost needed to glulam development.

The manufacturing process is now covered by BS 4169. Timber to be used in the process is kiln dried to a moisture content of 12 per cent. to 15 per cent. It is then graded. It must be borne in mind that defects are not so critical in glulam members because of the small cross sectional area of each laminate compared with the overall area of the finished member.

An appearance check is made, if appearance in the finished component is important. The laminae are then end jointed using scarf joints (1 : 12) or finger joints. Laminae are passed through a four cutter and a glue spreader. Glueing must take place immediately after planing, otherwise case-hardening commences and the effectiveness of the glue is reduced. The laminae are then clamped together in the jig so that the pressure is equally exerted over the complete length of the member. Pressure required is 0.7 N/mm^2 and this has to be maintained for 12 hours. The temperature and humidity control is very important during this time. The component then receives its finishing planing and then is painted or varnished.

Glues used are either urea formaldehyde or resorcinol formaldehyde, depending on whether the component is to be used internally or externally.

European Whitewood is excellent for the glulam process, but Douglas Fir is often used in North America. In the UK, Douglas Fir is less readily available and therefore is only used for components destined for severe conditions. Laminae for straight work are usually 30 mm planed to 25 mm, for curved work 28 mm.

Laminated beams have greater allowable stress limits than solid timber of a higher stress grade. The mean modulus of elasticity is always used, whereas in the case of solid members the minimum modulus of elasticity must be used. Therefore their stiffness is about twice that of solid members.

The advantages of glulam techniques are that curved forms in timber become possible, there is a greater dimensional freedom, less distortion and splitting and checking in the members, and the kiln drying of

the thin laminates used is quick and cheap compared with larger sections of wood.

Timber shell roofs are often associated with laminated beams (Fig.
3.11). This, like the mediaeval open timber roofs, is a wholly British
development. A timber shell roof has the advantage over a concrete
shell in its lightness (a 18.000 m square hyperbolic paraboloid in
timber would weigh 0.25 KN/m^2, whereas its equivalent in concrete
would weigh 1.5 KN/m^2). In addition it would not require expensive
formwork and has no curing time. The usual form adopted has been
that of a hyperbolic paraboloid, edge stiffened with laminated beams.

*Fig. 3.11   Laminated beams and timber shell roof of a Church.
Copyright: Rainham Timber Engineering Co. Ltd.*

In the diagram (Fig. 3.12) of a hyperbolic paraboloid, sections parallel
to A'B, BC', C'D and DA' are straight lines. Sections parallel to diagonal A^1C^1 are parabolic concave upwards and sections parallel to BD

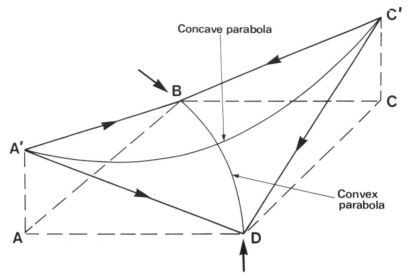

Fig.3.12 Geometry of H.P. shell. Sections parallel to edge members produce straight lines: sections parallel to diagonals produce parabolas: compression forces in edge members balanced by tie between B & D or buttresses at B & D.

are parabolic concave downwards. The lower corners of the shell need tying together or buttressing to avoid the roof spreading.

The shell itself is usually two layers of boarding with their grain running in opposite directions. Plywood can be used, or tongued and grooved boarding. The appearance required of the underside of the shell will determine the choice of material for the lower layer. It is recommended that boards used in the bottom layer should be finger jointed; while those in the top should be butt jointed. Shells have to have a minimum lift in order to avoid sagging or buckling. Their spans can range up to 18.000 m without difficulty and their erection is fast (a 155 m² hyperbolic paraboloid can be erected by nine men in 14 to 18 days).

The advantages of a timber shell roof construction are: low cost of materials, lightness (which means light foundations), speed of erection, the self decorated underside which merely needs treatment by clear varnish to complete, and a wide variety of achievable forms.

These three examples of modern timber engineering development serve to illustrate the degree of sophistication now achieved in timber technology. It has been suggested that in the future the starting point for any use of wood will be the pulp which is then formed into whatever shape is required. Maybe this is not the crazy idea it sounds. Certainly the trends are in that direction at present with greater emphasis on waste elimination and the utilisation of ever smaller sections of wood. However, light members will still be in their natural form, as timber's natural properties for simple members are difficult to improve upon.

Timber is now recognised as a valuable material and the use of all forms of its waste products are being investigated. Sawdust still presents a problem, although it can be formed into briquettes for burning as a domestic fuel. The Princes Risborough Laboratories of the Building Research Establishment are at present investigating the chipping saw, as opposed to the kerfed saw. This machine, as its name suggests, produces chippings and not sawdust. The chippings can be used in the production of particle board.

Planer waste is now used for particle board manufacture, although hardwood waste is not generally acceptable for this use.

Off-cut material can be either finger jointed to produce long lengths for less critical uses, or hogged for pulping or particle board.

This chapter started with the suggestion that the only common factor between carpentry and timber engineering was wood. In the last twenty years a revolution has occurred in the use of timber in construction. It has been a quiet revolution and one which has attracted little attention. Nevertheless, lessons learnt during the thousands of years of trial and error in timber's use, have been discarded in that time and timber has been looked at afresh in terms of science. This is not too extravagant a claim.

In the next chapters we are going to take a new look at wood, its properties and structural characteristics from the standpoint of the timber engineer.

4 The anatomy of wood

EVOLUTION

The evolution of the woody stem is closely related to climatic and geological changes and the first forests existed in the Middle Paleozoic period, 380 million years ago. The date at which wood was first used by man for structural purposes is a matter of conjecture, but it would be reasonable to assume that it has been in use as a structural material for at least 7000 years. (Early applications are described in Chapter 2).

In a broad sense the study of wood belongs to a branch of botany known as plant morphology. The cellular nature of plant structure was first observed by Robert Hooke (1636-1703) using an improved version of the microscope invented by a Dutch spectacle maker Zacharias Jansen in 1590. The cell walls, consisting of a number of concentric layers, are associated with the growth and strength of a woody stem. Although the walls of cells vary considerably in composition in different species the most important constituents are cellulose and lignin, making up 70 to 90 per cent of the wood tissue. The cellulose forms the fabric of the cell walls and has a high tensile strength, the lignin acting as a hardening agent.

THE WOODY STEM

Woody plants can be divided into two categories, firstly the cone bearing (Gymnospermae) and secondly the flower bearing (Angiospermae), into a subclass of which (Dicotyledoneae) falls trees with characteristics common to the oaks. To the first class belong trees such as pines, firs, spruces and hemlocks (conifers). In the vernacular, commercial timbers are divided into two groups softwoods and hardwoods. The distinction between hardwoods and softwoods is in fact a botanical one and the tables of strength properties given in Forest Products Research Bulletin No. 50 (1) indicate that many hardwoods are not as strong as some of the softwoods. For example English Willow (hardwood) has an average bending strength of under 70 N/mm^2 compared with about 90 N/mm^2 for Scots Pine (softwood). The process of development of a woody cell is gradual in which different cell types are formed (Fig. 4. 1). The primary tissues forming are the pith and the cortex (corky layer) the two being connected by pith rays. This is followed by the development of the cambium layer which produces the phloem (food conducting cells) and the xylem (water conducting and strengthening cells). The system consisting of trachieds and other cells forms the wood tissue. Tracheids are dead cells and form the bulk of the woody tissue of a stem. The commercial product, wood, is xylem. Annually the cambium forms a layer of new tissue many cells thick which appears as a series of concentric rings—annular rings (Fig. 4. 2). The portion of the ring formed in the early part of the

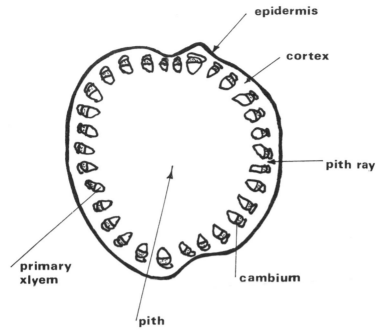

Fig.4.1 *Woody cell of a tree.*

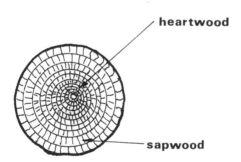

Fig.4.2 *Cross section through a tree.*

season is known as spring wood. The cell structure of the spring wood is radially larger and thinner walled than that grown in the summer. This difference in cell size is easily visible under relatively low powered magnification (Fig. 4. 3). A section through a woody stem reveals a central dark coloured zone (heartwood) and an outer lighter coloured zone (sapwood). The sapwood is gradually transformed into heartwood whose function is to provide mechanical support.

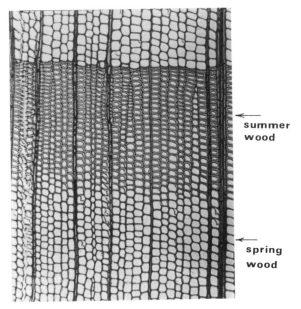

summer
wood

spring
wood

*Fig. 4.3 Photomicrograph (× 80) of woody stem of Douglas
Fir, TR section.*

THE STRUCTURAL CHARACTERISTICS OF WOOD

The tree has evolved over millions of years to produce a material of
great structural potential in terms of strength and stiffness, durability
and economy. The tree itself, is in effect, a cantilever beam—the trunk
a vertical cantilever, the branches cantilevering out from the trunk. In
tapering the trunk and branches, nature has provided a rational structural form, the greatest cross sectional area being provided at the base
of the trunk where the force actions are at a maximum (Fig. 4. 4). An
appreciation of the structural features of the woody stem can be obtained from an examination of sections through its principle axes TR,
LR and TL which are shown in Fig. 4. 5. An examination of a wedge of
material taken from the stem (Fig. 4. 6) indicates that timber should be
considered as an anisotropic material, that is, its strength will vary
according to the direction in which it is stressed. A cross section
through the stem TR reveals an orderly array of cells (trachieds)
which are aligned directly along the vertical axis of the trunk.

Figs. 4. 3 and 4. 7 are photomicrographs (×80) of woody stems, Douglas
Fir and European Whitewood respectively, the square or hexagonal
form of the cells being immediately apparent. The vertical lines on the
TR section are the wood rays which are revealed in cross section on
the TL section, Fig. 4. 8. The cell walls consist of a number of layers,
each of which contain microfibrils which are observable in an electron

Fig.4.4 The structural form of a tree.

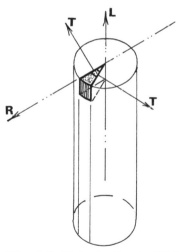

Fig.4.5 Principal axis TLR of the woody stem.

microscope. The principal constituent of the microfibrils is cellulose which gives the woody stem its structural strength.

Detailed information on the fine structure of wood is given in the first reference in the further reading list, page 202. The microstructure of timber is very complex but the TR section Figs. 4. 3 and 4. 7, indicates that a simplified structural model is obtained by considering the woody

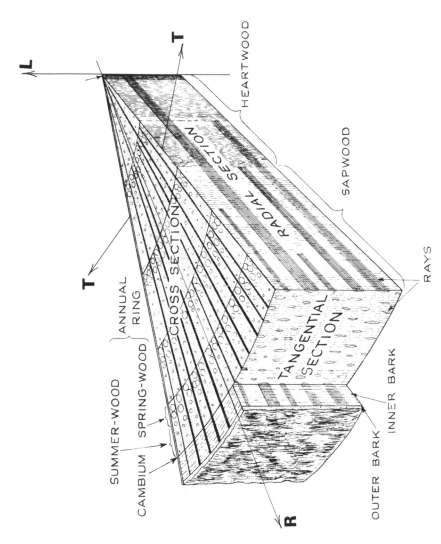

Fig. 4.6 The principal structural features of wood.

rays

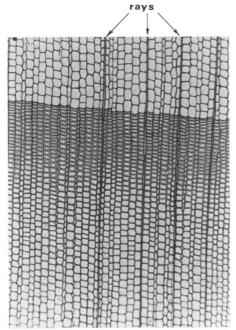

Fig. 4.7 Photomicrograph (× 80) of woody stem of Whitewood, TR section.

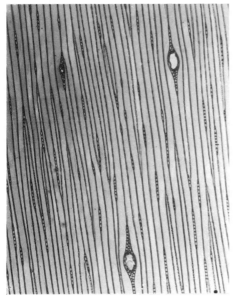

Fig. 4.8 Photomicrograph (× 80) of woody stem of Whitewood, TL section.

stem as a bundle of parallel tubes. Adopting this structural model some general conclusions can be made with regard to the resistance of wood to various force actions. Firstly there should be some correlation between strength and thickness of the cell walls which in turn should be related to the density of the material. Thus from an inspection of Figs. 4.3 and 4.7 it could be concluded that Douglas Fir is stronger than Redwood. This is confirmed in the Table 4.1 which shows the relationship between strength and density for various species. Factors such as growth defects and the percentage of moisture in the cell walls will, at this stage, be ignored and the values given in Table 4.1 are relevant to a moisture content of 12 per cent [1].

TABLE 4.1.   The relationship between strength and density for various species at a moisture content of 12 per cent

Species	Maximum bending strength N/mm^2	Maximum compression strength parallel to grain N/mm^2	Density Kg/m^3
Douglas Fir	93	52.1	545
Pitch Pine	107	56.1	769
Western Hemlock	83.0	47.4	465
Western Red Cedar	65.0	35.0	368
Baltic Redwood	83.0	45.0	481
European Spruce	72.0	36.5	417
European Larch	92.0	46.7	545

The above table indicates the general trend that the greater the density of the timber the higher the strength properties.

Returning to the structural model the effect of various force actions will now be examined. If it is loaded in a direction parallel to the tubes with a compressive force action (compression parallel to the grain) failure due to buckling of the cell walls would be expected and this is illustrated in Fig. 4.9. If the compressive force action is applied laterally (compression perpendicular to the grain) the cells will distort or crush quite easily and thus the strength in compression perpendicular to the grain is much less than that parallel to the grain.

In tension, the strength perpendicular to the grain is again very low but parallel to the grain it is much higher than in compression as there is no buckling. Considering the bundle of tubes acting as a beam Fig. 4.10 then the compression zone will act as a series of small columns restrained by the lightly stressed members nearer the neutral axis.

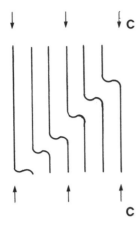

Fig.4.9 *Illustration of buckling of cell walls, due to compression parallel to the grain.*

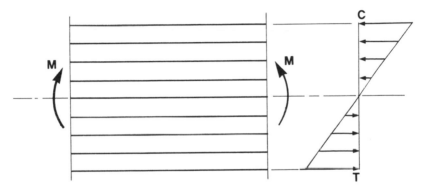

Fig.4.10 *Bundle of tubes acting as a beam.*

The degree of restraint offered by the lower fibres is greater with shallow beams as the rate at which the maximum fibre stress reduces is greater. This explains why the stress at failure for deep beams is somewhat lower than that for shallow beams.

To summarise, the bundle of tubes structural model gives a qualitative indication of the structural characteristics of wood, but its natural variability within the tree stem and from species to species necessitates the testing of a large number of specimens in order to evaluate strength properties for practical application. The following chapter deals with standard test procedures from which stress levels for structural design can be evaluated.

5 Strength tests on timber

Strength tests form the basis for assessing safe stress levels in structural timber. The tests are normally carried out on small specimens which are free from defects (e.g. knots and splits) and the period of load is only a few minutes. Thus the strength values obtained from the tests cannot be used directly for the design of structural elements. The test results must also be related to the amount of moisture present in the specimens (moisture content) as this will affect the values obtained (see Chapter 6). From a large number of specimens at a fixed moisture content the arithmetic mean of the strength values under a particular force action may be determined.

The arithmetic mean

$$\bar{x} = \frac{\Sigma x}{n} \qquad\qquad 5(i)$$

where x = measured strength value
    n = number of specimens

It would be very unwise to use the mean strength value as a basis for design as a large number of the specimens will have strength values below the mean. Thus it is necessary to obtain some measure of the degree of variation in strength values. It has been found that strength values of clear timber specimens tend to follow a distribution as shown in Fig. 5.1. This curve is symmetrical about the mean value $\bar{x}$. A

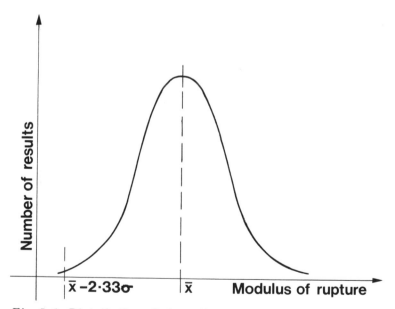

Fig. 5.1 *Distribution of strength values of clear timber specimens.*

measure of the dispersion of the results about the mean value, the standard deviation (2), is defined mathematically as follows:

$$\sigma = \left[\frac{\Sigma(x - \bar{x})^2}{n - 1} \right]^{1/2} \qquad \text{5(ii)}$$

The terms used in equation 5(ii) have been defined previously. It is now possible to estimate the value below which a certain percentage of the results will not fall. It can be shown that the following results apply:

68 per cent of the results lie within the range $\bar{x} \pm 1.0\ \sigma$
94 per cent of the results lie within the range $\bar{x} \pm 2.0\ \sigma$
98 per cent of the results lie within the range $\bar{x} \pm 2.33\ \sigma$

For design purposes it is necessary to decide what is an acceptable probability that a strength value will not fall below an estimated minimum. For timber a commonly accepted probability value is 1 in 100 (0.01). If the estimated minimum value is taken as $\bar{x} - 2.33\ \sigma$ then only 2 per cent of the results will lie outside the range $\bar{x} \pm 2.33\ \sigma$. For a symmetrical distribution curve the high and low values will be equally distributed and thus only 1 per cent of the values will fall below the the estimated minimum value. Thus there is a probability of 1/100 that the strength will fall below the estimated minimum value of $\bar{x} - 2.33\ \sigma$.

If the characteristic strength of timber is defined as the value below which only 1 per cent of the test results will fall it is given by

$$f_k = \bar{x} - 2.33\ \sigma \qquad \text{5(iii)}$$

BASIC STRESSES

The basic stress is defined as the stress which can be permanently sustained by a structural component of a particular species subjected to a particular force action, the component being assumed to consist of clear timber with no strength reducing characteristics. The strength of the timber specimen will be affected by its shape and size, the duration of loading and moisture content. Thus the characteristic strength must be divided by an appropriate factor which takes into account strength reducing characteristics and also the possibility of accidental overload.

The basic strength f_B is derived from the characteristic strength by dividing it by an appropriate factor of safety F.S.

Thus basic strength

$$f_B = \frac{f_k}{F.S.} = \frac{\bar{x} - 2.33\ \sigma}{F.S.} \qquad \text{5(iv)}$$

Basic strengths tabulated in the British Standard code of practice CP 112 : PART 2 : 1971 (3) are related to timbers having a moisture content not exceeding 18 per cent and those with a moisture content in excess of 18 per cent. For most strength properties a factor of safety of 2.25 is assumed to be appropriate. Some strength properties are considered below:

Bending

The specimens used by the Forest Products Research Laboratory (1) for the static bending test, from which the modulus of rupture is obtained, are 300 mm × 20 mm × 20 mm simply supported over a span of 280 mm. (Fig. 5.2). For a central load applied at a constant rate the load deflection diagram is recorded automatically up to failure.

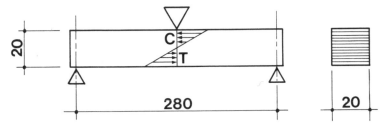

Fig. 5.2   Simple bending theory applied to a square section.

If it is assumed that simple bending theory applies then for a square section (Fig. 5.2),

$$\text{Bending moment} = \frac{WL}{4} = M = \frac{fd^3}{6}$$

$$\text{then } f = \frac{6M}{d^3} \qquad \qquad 5(v)$$

where f is defined as the modulus of rupture and is a measure of the ultimate bending strength of timber relevant to the size of specimen and loading conditions employed. Returning to the bundle of tubes structural model described in the previous chapter, it is unlikely that the stress distribution shown in Fig. 5.2 is relevant to timber as its strength in compression is less than that in tension due to the buckling of the tubes. Further, the relationship between stress and strain is not linear up to failure in tension or compression and thus the stress distribution curve is more likely to be of the form shown in Fig. 5.3. Another consideration is that the bending test is carried out on shallow specimens and for deeper specimens lower failure loads could be expected. Hence it must be emphasised that the modulus of rupture is only a measure of the bending strength of timber. The code of practice CP112:1971 (3) gives the following reduction factor to be applied to basic bending stresses to cover the deep beam effect for depths d (in mm) greater than 300.

$$\text{Reduction factor} = 0.81 \, \frac{d^2 + 92300}{d^2 + 56800} \qquad \qquad 5(vi)$$

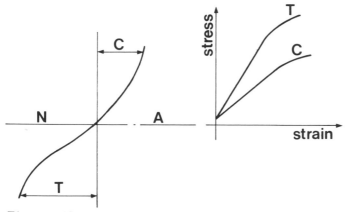

Fig. 5.3 Non-linear stress strain relationship for timber

For d = 300 mm the above factor is approximately unity and for d = 375 mm it reduces to 0.95. The modification factor is based on experimentalwork (4) on clear Douglas Fir sections up to 400 mm deep. Solid timber sections of depths greater than 300 mm are now rarely used (see Chapter 9) and thus the deep beam effect is not in general of great practical importance.

The results of static bending tests for various species of softwoods (1) are given in Table 5.1.

TABLE 5.1

	Moisture content per cent	Modulus of rupture N/mm³	Number of samples	Standard deviation σ
Douglas Fir (U.K.)	55	53	431	9.1
	12	91	425	16.9
Redwood (Baltic)	51	44	2628	7.9
	12.8	83	666	13.1
Spruce (European)	53	39	319	5.5
	13.8	72	324	10.2
Western Hemlock (Canada)	50	52	517	9.5
	12	97	237	17.5

Considering the Douglas Fir values in Table 5.1, the characteristic strength is given by

$$fk = 53 - 2.33 \times 9.1 = 31.8 \text{ N/mm}^2 \text{ (moisture content 55 per cent)}$$

$$fk = 91 - 2.33 \times 16.9 = 51.6 \text{ N/mm}^2 \text{ (moisture content 12 per cent)}$$

Adopting a factor of safety of 2.25 the basic stresses are as follows (see equation 5.iv)

$$f_B = \frac{31.8}{2.25} = 14.14 \text{ N/mm}^2 \text{ (for moisture content 55 per cent)}$$

$$f_B = \frac{51.6}{2.25} = 22.94 \text{ N/mm}^2 \text{ (moisture content 12 per cent)}$$

Shear

The shearing action on a timber section should be examined in relation to the principal axes TR, LR and TL. These axes are shown in Fig. 5.4, and for each plane it will be necessary to examine the effect of the complementary shears q on an elemental section Fig. 5.5 a, b, c. For the plane in which R is constant the stresses qLT will induce a sliding effect and the stresses qTL kinking.

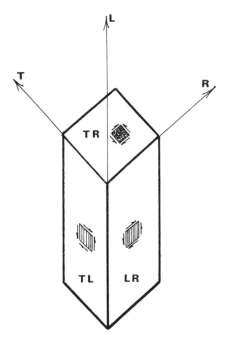

Fig. 5.4 Principal axes of a timber section.

In a similar manner for the plane in which T is constant qRL produces sliding and qLR kinking. Complementary shears qTR and qRT in the L constant plane will cause a rolling effect. Thus three types of shearing action should be considered—rolling, sliding and kinking. The rolling shear strength of timber is much lower than its strength in sliding or

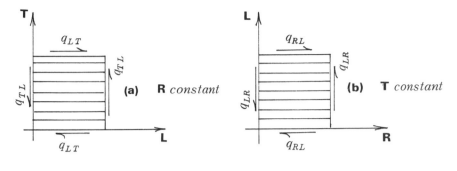

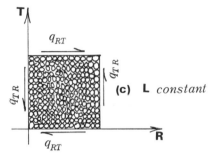

*Fig. 5.5 The effect of the complementary shear q on an
elemental section.*

kinking but fortunately it rarely occurs in practice with the exception of
plywood construction (see Chapter 9).

Shear tests are carried out on 20 mm cubes (1) in both the radial and
tangential plane on matched pairs of specimens and the results combined to give the average ultimate shear strength. The basic stress is
obtained from equation (iv) using a factor of safety of 2. 25. Some
typical results are given in Table 5. 2 (1).

Considering the Douglas Fir values in Table 5.2, the characteristic
shear strength is given by:

f_K = 6. 8 — 2. 33. 1. 28 — 3. 818 N/mm² (moisture content 55 per
   cent)

f_K = 11. 6 — 2. 33. 2. 29 = 6. 265 N/mm² (moisture content 12 per
   cent)

Adopting a factor of safety of 2. 25 the basic stresses are as follows
(see equation iv)

$$f_B = \frac{3.818}{2.25} = 1.70 \text{ N/mm}^2 \text{ (moisture content 55 per cent)}$$

$$f_B = \frac{6.265}{2.25} = 2.78 \text{ N/mm}^2 \text{ (moisture content 12 per cent)}$$

TABLE 5.2.

	Moisture content per cent	Shear strength parallel to grain N/mm^2	Number of samples	Standard deviation σ
Douglas Fir	55	6.8	764	1.28
(U.K.)	12	11.6	704	2.29
Redwood	51	5.9	2727	1.03
(Baltic)	12.8	11.3	690	1.69
Spruce	53	4.9	336	0.70
(European)	13.8	9.8	334	1.44
Western Hemlock	50	7.0	522	1.34
(Canadian)	12	14.1	244	2.67

Compression parallel to the grain

For obtaining the strength of timber in compression parallel to the grain specimens 60 mm high and 20 mm × 20 mm in cross section are used. The load is applied so that it is uniformly distributed over the cross section and the basic stress is obtained from equation 5(iv). A lower factor of safety is adopted for compression parallel to the grain (1.4) than for the bending test as:

(a)   the ratio of limit of proportionality to ultimate strength is about 25 per cent greater in compression parallel to the grain than it is in bending.
(b)   compressive strength is not affected to the same extent by the shape and size of the specimen.
(c)   defects influence compressive strength less than tensile. Some typical results are given in Table 5.3 (1).

Tension parallel to the grain

The tensile strength of timber is considerably influenced by the presence of defects but for clear specimens the tensile strength is greater than that in bending. It is considered adequate to adopt a basic stress equivalent to the bending value (see Table 5.1) rather than adopt a higher value with larger reductions for defects.

Compression perpendicular to the grain

Tests have shown that there is a strong correlation between the resistance offered to indentation by the side surface of timber and its compressive strength perpendicular to the grain. A measure of the resistance of timber to indentation (hardness) is obtained by using the Janka hardness tool and this is converted to a compressive strength perpendicular to the grain.

TABLE 5.3

	Moisture content per cent	Compressive strength parallel to grain	Number of samples	Standard deviation σ
Douglas Fir	55	24.6	1607	4.25
(U.K.)	12	848.3	1446	8.03
Redwood	51	21.0	2975	3.72
(Baltic)	12.8	45.0	690	7.63
Spruce	53	18.2	343	2.62
(European)	13.8	36.5	270	5.26
Western Hemlock	50	25.4	538	4.65
(Canada)	12	55.8	252	9.32

Modulus of elasticity

Values of modulus of elasticity are obtained from the static bending test described in section (a). For single members where deflection is important it is considered prudent to reduce the average value by 2.33 times the standard deviation. Some typical values are given in Table 5.4 (1).

TABLE 5.4.

	Moisture content per cent	Modulus of Elasticity N/mm^2	Number of samples	Standard deviation σ
Douglas Fir	55	8300	431	1850
(U.K.)	12	10500	425	2160
Redwood	51	7700	2603	1610
(Baltic)	12.8	10000	661	2040
Spruce	53	7400	304	1170
(European)	13.8	10200	324	2010
Western hemlock	50	8900	500	1440
(Canada)	12	11300	237	1970

The strength characteristics of commonly used structural softwoods given in tables 2-5 demonstrate the influence of moisture content on the test results. From these figures it is immediately apparent that if timber is to be used for structural purposes, then from strength considerations alone, moisture in timber is a characteristic of prime importance and is the subject of the next chapter.

6 Moisture in timber

The amount of water present in a sample of timber is generally expressed as a percentage of its oven dry weight and thus the moisture content may be defined as

$$\frac{\text{initial weight—final weight}}{\text{final (oven dry) weight}} \times 100 = \frac{\text{weight of moisture in timber}}{\text{oven dry weight of timber}} \times 100$$

With some timbers, when felled, (green) the weight of moisture is greater than that of the woody substance and thus the moisture content will exceed 100 per cent. Initially moisture is present in the cell walls and the cavities between them. In the first stages of drying out the cavity water is lost and there is a reduction in weight per unit volume of material. The volume however remains approximately constant until the moisture starts to leave the cell walls. The stage at which moisture is present in the cell walls only is termed the fibre saturation point—in the order of 30 per cent moisture content for commercial timbers. Below the fibre saturation point loss of moisture from the cell walls produces a volume reduction-shrinkage. This process is reversible and thus increase in moisture content up to the fibre saturation point will be accompanied by swelling. Before the material is put into service it is obviously desirable to ensure that the bulk of the total shrinkage has taken place and there are further important considerations:

(a) The strength of timber increases with reduction in moisture content below the fibre saturation value.
(b) timber may be regarded as practically immune from fungal attack at moisture contents below 20 per cent.

The moisture content of timber is affected considerably by the relative humidity of the atmosphere, that is, the amount of water vapour in the air expressed as a percentage of the amount it would hold if saturated. If a piece of green timber is exposed for a long period in dry atmosphere there will be a transfer of moisture from the timber to the air until a state of equilibrium is reached. The moisture content at which a stable condition occurs is known as the equilibrium moisture content. The equilibrium moisture content attained by timber in use will vary according to the relative humidity of the atmosphere. In a centrally heated building there is a small change in the relative humidity of the atmosphere throughout the year, the variation being from about 50 to 60 per cent. This corresponds to a moisture content variation in the wood from 10 to 12 per cent. Over 90 per cent of the softwood exported from Sweden to the United Kingdom is dried to a moisture content of 22 per cent or less before shipping; then assuming there is no further increase in moisture content up to the time of erection the shrinkage that will be associated with the loss in moisture content down to the equilibrium value will be in the order of 10 per cent. The change in geometry associated with reduction or increase in moisture content will depend on the species of the timber and the method by which it is sawn from the log. The sawing of timber from logs into suitable sizes for commercial use is referred to as conversion. In general the trunk of a tree can be sawn in one of two ways (Fig. 6.1):

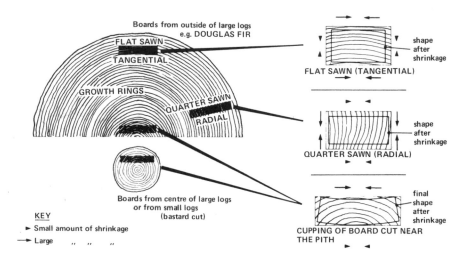

Fig. 6.1   Ways of sawing a tree trunk.

QUARTER SAWN (RADIAL)—the cuts are made in a radial direction roughly parallel with the rays across the growth rings from the bark towards the centre of the trunk

FLAT SAWN (TANGENTIAL)—the cuts are made roughly at right angles to the rays and tangential to the growth rings.

A uniformly dried piece of timber will not shrink appreciably until the fibre saturation point is reached, but differential shrinkage can take place at higher moisture contents due to surface drying and may lead to surface splitting and distortion of the timber. The shrinkage of timber from say 30 per cent moisture content will vary according to the species but will be at an approximately uniform rate and some average values are: 0.625 mm per 300 mm for each one per cent moisture content change in quarter sawn timber and 1 mm per 300 mm for each one per cent moisture content change in flat sawn timber. These values can lead to surprisingly large dimensional changes. If a flat sawn board 300 mm wide undergoes a moisture content change of say 18 per cent from the fibre saturation value (30 per cent) to an equilibrium value say 12 per cent, the corresponding reduction in width will be $1 \times 18 = 18$ mm. The shrinkage of timber in the longitudinal direction is small and generally less than 10 per cent of that in the tangential direction. The following values are quoted by the Swedish Timber Council for Redwood and Whitewood. (5)

	Redwood per cent	Whitewood per cent
Moisture content at fibre saturation point	26-28	30-34
During initial drying from the green condition to 12 per cent m.c. the shrinkage rate tangentially is	4.5 total	4 total
During initial drying from the green condition to 12 per cent m.c. the shrinkage rate radially is	3 total	2 total
During changes in moisture content subsequent to the initial drying, the movement tangentially expressed as a total for moisture change from fibre saturation point to 12 per cent m.c. is reduced to	2.2	1.5
During changes in moisture content subsequent to the initial drying, the movement radially expressed as a total for moisture change from fibre saturation point to 12 per cent m.c. is reduced to	1.0	0.7
In discussing permissible tolerances due to changes in moisture content, BS4471 states that actual sizes at moisture contents higher than 20 per cent up to 30 per cent, shall be greater by 1 per cent for every 5 per cent of m.c. and may be smaller by 1 per cent for every 5 per cent of m.c. below 20 per cent.		
The total longitudinal shrinkage on drying from the green condition to 12 per cent m.c. is	0.1-0.3	0.1-0.3

An explanation of why wood shrinks or swells far more across the grain than along the grain is given by N.A. de Bruyne (6). The cellulose molecules bunched together may be considered as a series of bricks forming a garden wall-like structure. The water creeps round the bricks but since there are many more gaps per unit length across the grain than along it, the amount of shrinkage or swelling per unit length is greater across than along the grain.

Methods of determining the moisture content of wood are as follows:

(a) The electric moisture content meter. This works on the principle that the drier the wood the greater the resistance to the passage of electric current. The range of reasonable accuracy is from about 6 to 25 per cent and as the electrical properties of all species of wood are not the same, correction values are required. Readings are taken by inserting prongs at the surface of a piece of wood which may be fairly dry compared with the centre. Fig.

6.2 illustrates the Protimeter Timbermaster which gives moisture content readings directly on a scale plate. It is calibrated for over 100 species of timber from about 7 per cent to above fibre saturation point. A hammer electrode is also available which enables the user to examine woods which are difficult to penetrate and also to obtain moisture readings in timber independent of surface moisture (Fig. 6.3). For laboratory work a more precise method is required.

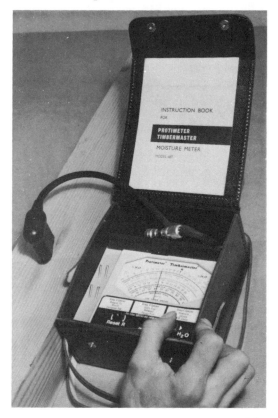

Fig. 6.2 A Protimeter Timbermaster Moisture Meter

(b) The oven drying method. A sample wood is first weighed, then dried for a period in an oven at about 100°C and again weighed. It is then replaced in the oven for a further period and another weighing made. This process is repeated until there is no loss in weight. The final (oven dry) weight is noted and the expression given at the beginning of the chapter can be used to evaluate the moisture content of the sample. As the ends of the timber board will be drier than the central portion it is advisable to cut samples away from the ends and weigh them with as little delay as possible.

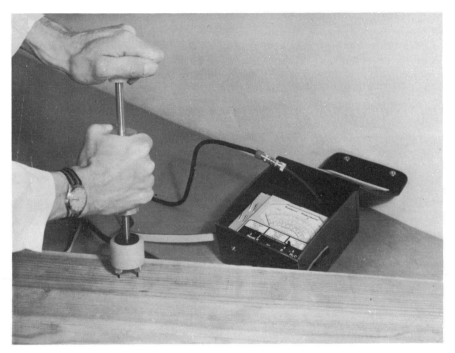

Fig.6.3 The Protimeter Timbermaster moisture meter being used with hammer-electrode for measuring moisture content at different depths in the wood

Conventional methods of drying timber prior to use for structural purposes are open air drying or in ventilated kilns. Thorough air drying is a slow process and if carried out in the United Kingdom will give final moisture contents in the range 18 to 22 per cent. Kiln seasoning gives much quicker drying and lower moisture contents can be obtained. Over 40 per cent of the production of Swedish softwood shipped to the United Kingdom is kilned in Sweden to an average of 18 per cent moisture content or less before shipping. Alternatives to the above are dehumidifiers, moisture extractions units and radio frequency, all of which are discussed in reference (7).

If timber is kiln dried or otherwise to a specified moisture content in a given situation, it is important to ensure that no significant change in moisture content takes place in transit or during storage on site. As wood is a hygroscopic material it will pick up moisture when exposed to damp air. Timber components, on leaving a factory production line at a moisture in the order of 12 per cent will pick up relatively little moisture during transit, say 24 hours, if closely stacked and covered with tarpaulins. If, however the components are left on site unprotected for a period of say 14 days during rainy weather the moisture content increase will be considerable. Thus the timing of deliveries of seasoned

timber components to sites is essential unless a suitable storage environment is available. Lack of care of seasoned timber defeats the
whole object of drying it to a specified moisture content and may lead
to distortion and splitting of the timber, stress concentrations at glue
lines and a reduction in strength of the cell structure of the wood.
The reduction in strength which accompanies an increase in moisture
content for a given species is immediately apparent from Fig. 6.4.

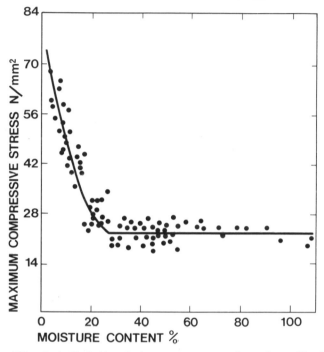

*Fig. 6.4 Relation between compressive strength of Scots Pine
and its moisture content.*

The curve shows the relationship between compression strength and
moisture content of clear specimens. If results are plotted directly
from tests there is a wide dispersal of the results due to the individual
characteristics of the test specimens. As indicated in Chapter 4 the
strength of timber can be related to the amount of woody substance per
unit volume. As this can vary from specimen to specimen of the same
species it is necessary to correct the results so that the strength
values will correspond to those for specimens with the same wood
content.

Increase in strength with reduction in moisture content follows a similar trend for most strength properties but at low moisture contents the
wood tends to be more brittle. Above the fibre saturation point the
strength properties remain at approximately constant and minimum
values. In CP112: 1971 The Structural Use of Timber (3) effect of

moisture content on strength properties is recognised by giving basic stresses for two conditions:

(a) Green stresses which apply to timber having a moisture content exceeding 18 per cent

(b) Any stresses which apply to timber having a moisture content not exceeding 18 per cent.

A further consideration in the evaluation of strength properties of timber is the presence of growth defects and this will be the subject of the next chapter.

7 Grading of timber

In Chapter 5 it was shown that basic strength values for timber are determined on a statistical basis from tests on clear specimens. The use of clear timber for structural applications is not in general commercially viable. Thus it is necessary to modify equation 5(iv) to allow for natural features which have the effect of reducing the strength of the member containing them. In the United Kingdom it has been normal practice to grade timber according to the type, size and location of defect present. This process is carried out visually to give four strength ratios—40, 50, 65 and 75 per cent. The strength ratio is defined as the ratio of the grade stress to the basic stress. Using the symbol K_G to denote this ratio, equation 5(iv) can be modified to give the grade strength f_B^G as follows

$$f_B^G = \frac{\bar{x} - 2.33\sigma}{F.S.} \cdot K_G \qquad\qquad 7(i)$$

K_G has four values 0.4, 0.5, 0.65 and 0.75. Thus 40 grade timber has a strength of only 40 per cent of the basic value.

The natural features which influence the strength of timber are as follows

(a) slope of the grain—can vary from 1 in 6 to 1 in 14 according to grade
(b) fissures—separation of fibres in various forms, checks, shakes and splits
(c) knots—these produce disturbance of the grain which causes strength reduction
(d) wane—the rounded part of the tree section appearing on the corner of a converted piece of timber.

The effect of these defects on the strength of timber is depedent on their type, size and location within the section and the force actions being considered. Rules for the visual grading of timber are given in reference (3) and further useful guidance is given in a TRADA pocket book (8). The application of these rules to every piece of timber arriving at the stockyard of a factory producing structural components would be a monumental task involving a number of technicians qualified in visual grading. In practice it is unlikely that these rules are rigidly applied and reliance is made on the fact that much of the timber imported from Europe and North American has a strength ratio of at least 50 per cent. The main visual grade used in structural components is probably the 50 grade and thus if the supplier is known to the designer there should be little difficulty meeting the 50 grade requirement. Unfortunately rules for grading timber vary from country to country and this leads to confusion. However useful guidance is available from the Swedish Timber Council (9) and the Council of Forest Industries of British Columbia (CFI) (10). In Sweden there are six basic qualities designated I, II, III, IV, V and VI, which apply equally to Redwood and Whitewood. The system normally employed at the sawmills is not to separate the I to IV basic qualities but to sell them as a grouped grade known as 'unsorted' quality. Fig. 7.1 illustrates typical basic qualities and is reproduced by permission of the Swedish Timber Council.

The laborious process of visual grading and the increasing demand by engineers for design data for timber to be of the same form as for other materials has led to the development of a machine for stress grading timber. Further it must be recognised that the stress values for visually graded timber are conservative for many of the pieces within a grade as the grade stresses are based on the assumption that the visible defects are associated with material of minimum strength that is the mean value minus 2.33 times the standard deviation as indicated in equation 7(i). From a visual inspection of a piece of timber it is not possible to assess the density of woody substance which is subject to considerable variation within a tree and from tree to tree. Therefore it is possible for a piece of timber containing knots to be stronger than another piece which is free from visible defects. The development of a machine for grading timber was the result of the realisation of the existence of a statistical relationship (11) between modulus of rupture and modulus of elasticity for clear specimens of South Australian plantation grown pine (Pinus, Radiata, Kenya). Subsequently investigations in America established a significant correlation between bending stress and modulus of elasticity for timber containing defects. Thus a grading machine would be required to sense continuously the flexural rigidity of a piece of timber. It would have the advantages of automation and take into account the defects and natural variation in strength due to the changes in density of the woody substance within each piece. Investigations carried out by the Forest Products Research Laboratory, England led to the following conclusions (12),

(a) The relations between modulus of rupture as a joist and modulus of elasticity yielded correlation coefficients better than 0.7
(b) the relations were equally good, irrespective of whether the modulus of elasticity was measured as a joist or a board.
(c) modulus of elasticity gives a better index of bending strength than does a measure of knot size or knot area ratio, the factors on which visual stress grading are principally based.

These conclusions justified further work on the assumption that the machine would sense continuously or at discrete intervals the deflection of a piece of timber being fed through it. The maximum deflection recorded can be used to obtain a grade stress from the modulus of rupture/modulus of elasticity relationship. It was found that this relationship was not significantly influenced by the level of stress in the timber during the deflection sensing up to 12.5 N/mm^2 and the influence of the span over which the deflection is measured was investigated. Further work led to the following general conclusions (12)

(a) Changes in moisture content have no significant influence on the stiffness (EI) of timber. As the moisture content decreases,

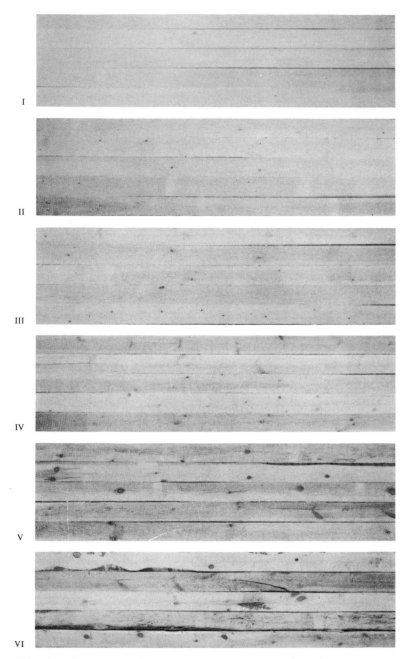

I

II

III

IV

V

VI

Fig. 7.1 The look of typical basic qualities of timber.

modulus of elasticity increases, but this is offset by a reduction in the second moment of area of the section caused by shrinkage.
(b) The same regression line for strength on modulus of elasticity may be used for a number of different species of timber.
(c) There are acceptable correlations for compression and tension strength and modulus of elasticity as a board, but not for shear strength.
(d) Size of section influences strength—the larger the section the lower the apparent ultimate stress, but with no corresponding reduction in modulus of elasticity. The basic relations used for control setting of a machine should therefore be based on a standard size of section and appropriate modifications made to the corresponding grade stress values to correct for the effect of size.
(e) Different regression equations apply for forecasts of bending strength as a board and as a joist.

Commercial grading machines are now available, manufactured in America, Australia and the United Kingdom. The Australian Computer-matic machine is considered to be the most suitable for European conditions as it can accept timber with sawn surfaces. A Computer-matic machine has been obtained by the Forest Products Research Laboratory. It was not possible to identify in a single pass the four grade ratios 75, 60, 50 and 40 as the difference between them is too small. However it has been possible to establish interim stress values for European Redwood and Whitewood (13). The machine grades have been designated M50 and M75 to distinguish them from the visual grades. The bending and shear stress values for these grades are the same as for the 50 and 75 visual grades specified in CP112: 1967. The modulus of elasticity values are higher and this is a marked advantage as deflection is generally the governing factor in the design of solid timber joists for domestic usage.

Finally, a comparison of yield and grade stresses for mechanically and visually stress graded European Redwood and Whitewood will be made. Table 7.1 (12) shows that grading by machine gives much better yields in the higher grades, or, if the same yields were maintained as for visual grading a substantial increase in the grade stress values would be justified.

TABLE 7.1

Grade Classification	Grade Stress N/mm²	Per cent yield visual grading	Per cent yield machine grading	Grade stress for machine grading giving same yield as visual grading N/mm²
75	10.0	35	68	11.7
65	8.6	33	18.5	10.0
50	6.5	23	12	7.9
<50	—	9	1.5	—

From Table 7. 1 the advantages of machine grading are immediately
apparent and a further point to be appreciated is that there is a lack
of technicians trained in carrying out the visual grading rules and thus
visual grading is generally applied only in a nominal manner in practice. In general timber imported from Scandinavia and North America
can be relied upon to meet the 50 grade requirements but this is hardly
a satisfactory basis for design. It should also be noted that visual
stress grading does not take into account the most important strength
criterion, that is the fibre structure of the material.

The new British Standards Institution specification 'Timber Grades for
Structural Use' (BS4973) was issued in 1973 and specifies two methods
of grading timber for structural use, namely, visual stress grading and
machine stress grading. The standard specifies two visual grades,
General Structural grade (GS) and a higher grade, Special Structural
grade (SS) and the principle of the knot area ratio (KAR) is adopted as
a means of determining the maximum permissible knots for a given
grade. The knot area ratio is defined as the ratio of the sum of the
projected areas of all knots at a cross section to the cross sectional
area of the piece. The two machine grades specified, which may be
used as substitutes for the two visual grades, are MGS for GS and MSS
for SS. Stress levels for the four grades (GS, SS, MGS, and MSS) are
given in the 1973 amendment to CP112: 1971: Part 2 (metric units).
Until machine graded timber is generally available the two approached
to strength grading will be required, but eventually machine grading
should replace visual grading due to its reliability and ultimate cost
benefits.

8 Long term loading

It was shown in Chapter 5 that the basic stresses for timber are obtained on a statistical basis from short term tests on clear specimens. These stresses are then modified to allow for natural defects as outlined in Chapter 7. In practice a timber structure or component may be required to sustain a load representing a significant proportion of its short term ultimate load throughout its useful life, which could be in excess of 50 years. It is generally recognised that:

(a) the strength of a timber member loaded continuously for a number of years is slightly in excess of half that required to cause failure under short term loads.

(b) under sustained loading the deformation of timber increases with time and then stabilises or produces failure of the member depending on the magnitude of the initial stress.

The factor of safety, generally 2.25, used to convert the characteristic strength to a basic strength (equation 5(iv)) includes an allowance for the effect of long term loading. Statements (a) and (b) above are of great importance to designers and duration of loading is covered in some detail in the commentary on the British Standard Code of Practice CP112:1967 by L. G. Booth and P. O. Reece (4). As the reduction factor of 2.25 is based on the assumption that all the loading is long term it would seem reasonable that some adjustment could be made for loading of short duration. In CP112:1971 the following modification factors are given for application to basic and grade stresses (3).

Duration of loading	Modification factor K_D
Long term (e.g. dead + permanent imposed)	1.0
Medium term (e.g. dead + snow load + temporary loads)	1.25
Short term (e.g. dead + imposed + wind, dead + imposed + snow + wind)	1.50

Using the modification factors K_G (grade) and K_D (duration of load) the following expression may be used to obtain the permissible stress f_p for a particular species of timber

$$f_p = \frac{x - 2.33\sigma}{\text{F.S.}(2.25)} \cdot K_G \cdot K_D \qquad \text{8(i)}$$

As with concrete structures it is necessary to consider deflections under sustained loading in the design of timber structures. The total deflection at any time t is the sum of the initial elastic deformation W_e and the creep deformation W_c. The creep deformation is the additional deformation which occurs with time and the creep factor ϕ may be defined as the ratio

$$\frac{\text{creep deformation}}{\text{initial deformation}} = \frac{W_c}{W_e} = \phi$$

The total deformation at a time interval t after the initial elastic deformation is thus

$$W_t = W_e + W_c$$
$$= W_e + \phi W_e$$
$$= W_e (1 + \phi) \qquad\qquad 8(ii)$$

Under normal stress levels the deformation/time curve should take the form OA as shown in Fig. 8.1. Thus after an interval of time the rate of change of creep deformation should decrease and at point A the curve should be approximately parallel to the time axis. At higher stress levels the curve may take the form OB that is the rate at change of creep increases with time. For the stress levels implicit in equation 8(i) the deformation time curve will take the form OA. Curve OB indicates the danger of using high design stresses for sustained loading conditions. Research work on the deformation time behaviour of timber is discussed in reference (4) and the results are dependent on the stress level (that is the ratio of the applied stress to the ultimate stress expressed as a percentage), moisture content, change in moisture content and temperature. A plot of creep factor ϕ against time for various stress levels will take the general form shown in Fig. 8.2 for a given moisture content and temperature.

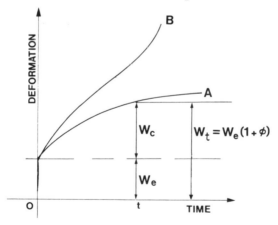

Fig. 8.1   Deformation/time curves for timber

Of particular importance is change in moisture content due to atmospheric conditions for which the creep factor ϕ may be in the order of 1.25 for preseasoned beams. For beams seasoning under load ϕ may reached a value of 7.0. A creep time curve for air dry beams of

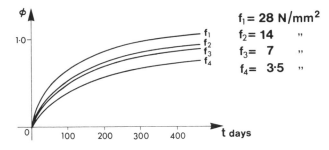

Fig. 8.2 *A plot of creep factor φ against time for various stress levels.*

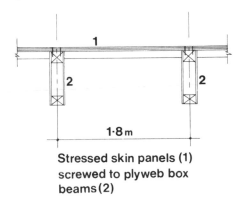

Stressed skin panels (1)
screwed to plyweb box
beams (2)

Fig. 8.3 Roof beams spaced at 1.8 m. cross centres.

mountain ash at 40°C given in (4) indicates a value of ϕ in the order of 0.75 at 480 days at a stress in the range 7 to 14 N/mm^2.

Let us consider a typical design problem in which it is necessary to estimate the deformation due to sustained loading. Roof beams spaced at 1.8 m centres carry the following loads per square metre over simply supported spans of 9.0 m (Fig. 8.3)

imposed (snow) loading = 0.75 KN/m^2
dead loading = 0.5 KN/m^2

Thus the loadings per metre run of beam are

imposed   1.8 × 0.75 = 1.35 KN/m

dead $\begin{bmatrix} \text{timber} \\ + \\ \text{roofing} \\ + \\ \text{ceiling} \end{bmatrix}$ 1.8 × 0.5 = 0.90 KN/m

The form of construction is such that the beams act compositely with a stressed skin deck to carry the imposed loading. The self weight of the roof structure is carried by the beams only which are of box section. The self weight of the roof structure excluding the roofing felt and ceiling is approximately 60 per cent of the total dead load, that is, $0.6 \times 0.9 = 0.54$ KN/m. Under the loading of 0.54 KN/m the elastic deflection of the beam is estimated to be 11.5 mm. For the remaining 40 per cent of the dead load composite action may be assumed (see Chapter 9) and the estimated deflection is 6.0 mm. Thus the total deflection under dead load is $11.5 + 6.0 = 17.5$ mm. Under dead load the stress level is low and a creep factor ϕ in the order of 0.4 may be expected. This will result in a total long term deflection under dead load of (equation 8(ii))

$$W_t{}^D = W_e^D \ (1 + \phi)$$
$$= 17.5 \ (1 + 0.4)$$
$$= 17.5 \times 1.4$$
$$= 24.5 \text{ mm}$$

Finally an estimate of the deflection due to the imposed (snow) loading is required. Records dealing with the occurrance of snow are maintained at the Meteorological Office at Kew Observatory. Unfortunately these records do not indicate the depth of snow but only the number of days on which snow was observed to be lying on the ground. The design snow load represents a depth in the order of 0.6 m and it has been deduced (4) that this loading would be borne by a structure for 30 days in 50 years. Thus it would be reasonable to assume that the creep factor ϕ under imposed loading will be small, say 0.2. An estimate of the elastic deflection due to the imposed loading is 22.5 mm giving a total deflection

$$W_t{}^1 = 22.5 \ (1 + 0.2)$$
$$= 27.0 \text{ mm}$$

The total deflection (long term) under dead and imposed loading is

$$W_t{}^D + W_t{}^1 = 24.5 + 27.0$$
$$= 51.5 \text{ mm}$$

This result will give an indication of the camber required to ensure adequate roof drainage. An estimate of stress levels required for a complete design has not been included as full details of calculation procedures for structural elements are the subject of Chapter 9.

9 Design of structural elements

INTRODUCTION

The conventional approach to the design of structural elements in timber, concrete, steel and masonry has been to adopt the elastic theory
for the evaluation of force actions and to determine stress levels. The
stress levels at design load (permissible stresses) are limited to a
fraction of the ultimate value for the material to ensure there is an
adequate margin of safety against failure. As the common materials
of construction do not obey Hookes' Law up to failure, this margin of
safety is merely the ratio of the ultimate to the working load stress and
does not give the true factor of safety against failure. In the past
twenty years the load factor method of proportioning structional elements has been used for steel and reinforced concrete, the load factor
being defined as the ratio

$$\frac{\text{Ultimate load}}{\text{Working load}}$$

This approach has not as yet found practical application in the design
of timber structures. Considering a section subjected to bending typical stress distributions at ultimate load are shown in Fig. 9.1. An in-
vestigation into ultimate beam theory for rectangular timber beams

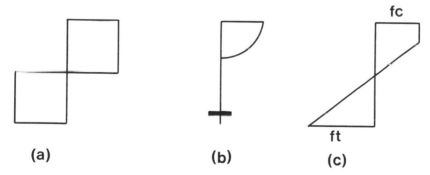

(a)	(b)	(c)

Fig. 9.1 Typical stress distributions for flexure at ultimate
conditions: (a) steel, (b) reinforced concrete, (c) timber.

has been carried out by D.N. NWOKOYE* and the stress distribution
shown in Fig. 9.1 (c) is proposed. The section analysis is then carried
out using the following assumptions

(a) the modulus of elasticity in compression is the same as in tension

(b) wood in the compression zone behaves as an ideal elastic-plastic

---

* Nwokoye, O. N. An introduction into an ultimate beam theory for
rectangular timber beams—solid and laminated. Timber Research &
Development Association, April 1972

material, whilst in the tension zone it behaves as a linearly elastic
material up to failure

(c) plane sections before bending remain plane after bending

(d) the bending strain across the depth varies linearly until the proportional limit is reached

(e) the extreme fibre stress at the proportional limit in bending equals
the ultimate compressive strength parallel to the grain

Thus if fc and ft (Fig. 9.1 (c)) are known then from considerations of
equilibrium and the geometry of the stress diagram an expression for
the ultimate strength of the section may be obtained.

In recent years the idea of referring to a single failure criterion has
been replaced by the comprehensive concept of limit states in which the
local or overall behaviour of the structure is considered at all stages
of loading. A limit state is defined as a state of unfitness for use and
of primary importance is the limit state of collapse. It is also necessary to consider serviceability limit states such as deflection and
vibration. Partial safety factors are introduced for each limit state and
consist of reduction factors for mechanical strength (γm) and enhance-
ment factors for loading (γ_L). The terms characteristic strength and
characteristic load are also introduced.

The characteristic strength of a material may be defined as the
strength that is normally expected to be exceeded and can be defined
statistically as follows:

$$U_K = U_m - K.\sigma,$$

where U_K = characteristic strength,
  U_m = mean strength,
  σ = standard deviation, and
  K = factor to ensure that the probability of the characteristic
  strength not being exceeded is small.

For timber the value of K is at present taken as 2.33 (see Chapter 5)
The design strength U_D may be expressed as

$$U_D = \frac{U_K}{\gamma_m}$$

The characteristic load may be defined as the load which is not likely
to be exceeded during the useful life of the structure. At present the
characteristic load cannot be defined statistically and it is assumed
that the characteristic load is the load given in CP3: chapter V, part 1:
dead and imposed load 1967. The design load is obtained by multiplying the characteristic load by γ_L, the appropriate partial safety factor for the limit state being considered.

Thus the design load may be expressed as

$$W_D = W_K.\gamma_L$$

where W_K is the characteristic load.

The latest edition of the code of practice for structural concrete (CP 110 : 1972) is written in limit state terms and work is in hand for similar documents for bridges, steel and timber. The present edition of the code of practice for structural timber (CP 112: part 2: 1971 metric units) is based on traditional methods and forms the basis of the approach to the design of structural elements given in this chapter.

In Chapter 8 the following expression was developed (equation 8(i)) to obtain the permissible stress for a particular species of timber, that is

$$f_p = \frac{\bar{x} - 2.33\sigma}{F.S.} \cdot K_G \cdot K_D$$

where K_G and K_D are modification factors for the timber grade (visual) and the duration of loading. If the timber is machine graded M50, M75 (13) the bending and shear stresses are the same as for the 50 and 75 visual grades and these values are represented by the portion

$\dfrac{\bar{x} - 2.33\sigma}{F.S.} \cdot K_G$ of the above equation.

For machine graded timber a factor of safety of 3.0 is recommended but it should be noted that this value embraces the presence of defects.

Equation 8(i) will be used as a starting point for structural calculations and the value of 2.33 ensures that the characteristic strength $\bar{x} - 2.33\sigma$ has a probability of being exceeded ninety-nine times in a hundred in test. This factor would appear to be unnecessarily severe if compared with the current practice adopted in the design of concrete structures where it is considered adequate to use a characteristic strength of $\bar{x} - 1.6\sigma$. This increases the probability of the characteristic value being exceeded to 1 in 20. Values of mean ultimate strength $\bar{x}$ and the standard deviation σ may be obtained from Forest Products Research Bulletin No. 50 (1) for over 200 home grown and imported softwoods and hardwoods in the green and air dry conditions. Alternatively grade stresses may be obtained directly from CP112: 1971 for three groups of species S1, S2 and S3. These groups are listed below, Table 9.1.

TABLE 9.1

Group	Timber	
S1	Douglas Fir	Imported
	Pitch Fir	Imported
	Douglas Fir	Home-grown
	Larch	Home-grown
S2	Western Hemlock (unmixed)	Imported
	Western Hemlock (commercial)	Imported
	Panama Pine	Imported
	Redwood	Imported
	Whitewood	Imported
	Canadian Spruce	Imported
	Scots Pine	Home-grown

TABLE 9.1 (continued)

Group   Timber

S3	European Spruce	Home-grown
	Sitka Spruce	Home-grown
	Western Red Cedar	Imported

It is not always possible to specify a particular species and thus soft-wood species are divided into three strength groups. The grouping of species sometimes results in a less efficient use of the material than would be obtained with the use of a particular species at its appropriate stress level. The properties of Swedish Redwood and Whitewood have been published by the Swedish Timber Council (14), those for Pacific Coast Hemlock by CFI (15) and data for hardwoods is available from the Timber Research and Development Association (16).

The versatility of timber as a structural material has been demonstrated in previous chapters and having obtained a basic relationship from which safe stress levels may be derived it is now possible to consider the procedure for proportioning structural elements. It is not intended to cover all structural applications in detail but simply to give a broad outline of calculation techniques relating to strength and serviceability for a number of commonly used elements starting from the simple beam.

THE SOLID BEAM

The rectangular solid section, a refinement from the log via the development of mechanised sawing techniques is still very widely used as a structural element for floors and roofs of relatively short span. It is important to realise the limitations on section sizes normally available—although the trees from which sawn sections are derived can grow to a height of 54 m maximum lengths obtained in the U.K. have an upper limit of about 10 m. Typical maximum spans and section sizes are listed below.

	Length m	Width mm	Depth mm
Swedish Redwood and Whitewood	5.7	75	225
Pacific Coast Hemlock	7.2	44	300

Lengths up to 12.0 m can be obtained from Sweden by means of finger-jointing. Larger section sizes and greater jointless lengths are more readily available from North America than Europe. Alternatively hardwoods can be specified as a means of achieving non standard spans. However normal domestic requirements are for spans up to about 5 m.

It is common practice to design timber joists as simply supported beams at fairly close centres (300-600 mm) assuming no interaction between the joists and the decking. In practice some interaction will be achieved, even with nominal nailing, and CP112: 1971 recognises this by allowing a 10 per cent increase in grade stress where 4 or more members can be considered to act together to support a common load and the mean value of the modulus of elasticity may be used to calculate deflections. Force actions normally considered in the design of simple joists are bending, shear and end bearing. A deflection check is also required to ensure adequate serviceability, that is avoidance of damage to surfacing, ceilings and partitions. For uncambered beams the deflection under dead plus superimposed load is commonly limited to $0.003 \times$ span. Finally it is necessary to ensure that the member has adequate lateral stability as it is possible for a beam to buckle laterally at relatively low loads if it is not restrained in position at the ends and held in line by decking and bridging or blocking pieces. CP112: 1971 gives limiting depth to breadth ratios for rectangular sections with various degrees of lateral support. The structural calculations is a confirmation process relating to an initial estimate of self weight and imposed loads. The self weight of the joists, decking and ceiling will rarely exceed 0.5 KN/M^2 and this can be used as a preliminary design figure for domestic floors, or alternatively, separate figures can be allocated to each part of the floor construction based on tabulated data. The joist spacing x (Fig. 9.2) will be in the range 0.30 to 0.60 m for a span Lm and total loading per square metre (dead and imposed) of W^{KN}/m^2, the load carried by the joists is given by

$$W = w.x.L \text{ KN} \qquad 9(i)$$

Starting with bending, design equations can be developed.

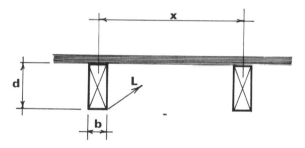

Fig. 9.2 Solid timber floor joists at spacing x.

Bending

The applied bending moment $M = \dfrac{WL}{8}$ KNm. For a permissible bending stress f_pN/mm^2, depth dmm, breadth bmm the resistance moment of section

$$= f_p \times \frac{bd^2}{6}$$

thus $\dfrac{WL}{8} \times 10^6 = f_p \times \dfrac{bd^2}{6}$

$$W = \frac{4}{3} \times f_p \times \frac{bd^2}{L} \times 10^{-6} \qquad 9(ii)$$

The allowable total load W is related to f_p, b, d and L and in general the depth d can be determined from equation 9(ii) for an assumed value of b, usually in the range 25 to 75 mm. Having established the section dimensions these can be checked for shear.

Shear

For a rectangular section the distribution of shear stress is parabolic and the maximum value is $1.5 \times$ the mean value. The maximum shear force is half the total load

$$= \frac{W}{2}$$

thus $q_p = \dfrac{W}{2bd} \times 1.5 \times 10^3$

where q_p = permissible shear stress parallel to the grain $^N/mm^2$

or $W = \dfrac{4}{3} \times q_p \times bd \times 10^{-3}$ 9(iii)

The section will be adequate for shear if the value of W obtained from equation 9(iii) by substituting the b and d values obtained from the bending equation is greater than the allowable load for bending that is

W 9(iii) $\geqslant$ W 9(ii)

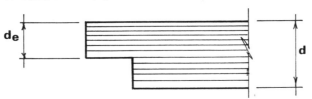

Fig. 9.3 Beams are notched at the ends.

It is common practice to notch beams at the ends Fig. 9.3 which reduces the depth of section resisting shear to de and causes a concentration of stress. This is allowed for in CP112: 1967 by calculating the shear strength using an effective depth de and applying a reduction factor de/d. Thus equation 9(iii) will take the form

$$W = \frac{4}{3} \times q_p \times b \times de \times \frac{de}{d} \times 10^{-3} \qquad 9(iv)$$

Bearing

Where the joist sits on its end supports the timber will be in compression perpendicular to the grain Fig. 9.4. Thus the allowable load in bearing is the bearing area multiplied by the permissible bearing stress. The bearing reaction is W/2 thus

$$\frac{W}{2} = b \times t \times np \qquad 9(v)$$

$$W = 2 \times b \times t \times np$$

where np = permissible bearing stress $^{N}/mm^2$

again W 9(v) ⩾ W 9(ii)

Where the bearing is located 75 mm or more from the ends of a beam and the bearing length is less than 150 mm CP112: 1971 allows a stress increase as listed in the table below

Length of bearing (mm)	10	15	25	40	50	75	100	150 or more
Value of modification factor applied to compression perpendicular to grain	1.74	1.67	1.53	1.33	1.20	1.14	1.10	1.00

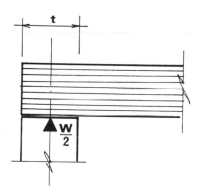

Fig.9.4 Joist sitting on end support.

Deflection

With timber joists deflection is often the critical design consideration especially if the minimum modulus of elasticity is adopted (mean − 2.33σ). However for load sharing systems the mean value is more appropriate and for a total load W the deflection δ is given by

$$\delta = \frac{5}{384} \times \frac{WL^3}{EI}$$

where E = Modulus of elasticity $^{N}/mm^2$

  I = 2nd moment of area mm⁴ = $\frac{bd^3}{12}$

If the deflection is limited to 0.003 × span then

$$0.003 \times L = \frac{5}{384} \times \frac{WL^3}{EI}$$

For the above to be dimensionally correct

$$0.003 \times L \times 10^3 = \frac{5}{384} \times \frac{W \times 10^3 \times L^3 \times 10^9}{E \times I}$$

or

$$W = 0.0192 \times \frac{Edb^3}{L^2} \times 10^{-9} \qquad\qquad 9(vi)$$

The above considers bending deflection only. Deflection due to shear is generally ignored in the design of solid timber joists but will be considered in subsequent sections of this chapter:

again W 9(vi) ⩾ W 9 (ii)

The arithmetic work involved in designing solid timber joists can be minimised by using the excellent design charts prepared by TRADA (17). The authors have found these charts particularly useful for the design of domestic floors and roofs. The equations developed above apply to uniform loading but a similar procedure may be adopted for other forms of loading. A few calculations will soon demonstrate the span and load limitations of solid timber beams and thus we can now examine alternatives to the solid timber beams which will increase the span range and load capacity.

PLY WEB BEAMS

Ply web beams are widely used for spans in excess of those for which solid timber is suitable and have the structural advantage that relatively small solid timber sections can be used for the top and bottom chords which are spaced well apart by the plywood web material Fig.

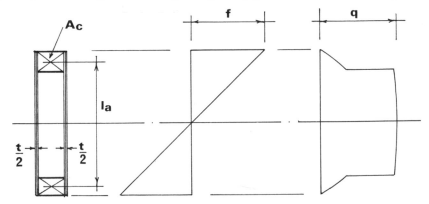

Fig.9.5 Ply web beams: solid timber is placed at positions of maximum bending stress.

9.5. Thus the solid timber is placed at the positions of maximum bending stress and is working at its greatest efficiency. From the bending and shear stress distributions in I or box sections Fig. 9.5 it can be seen that for preliminary design the following formulae are appropriate (notation as in Fig. 9.5).

$$\text{Chord area } A_c = \frac{M}{la \times f} \qquad\qquad 9\text{(vii)}$$

$$\text{Web thickness } t = \frac{Q}{la \times q} \qquad\qquad 9\text{(viii)}$$

where   M is the applied bending moment
   Q is the applied shear force
   f is the permissible chord compressive stress
   q is the permissible web shear stress
   la is the distance between the centroids of top and bottom
     chords.

Equations 9(vii) and (viii) may be used as a basis for estimating a trial chord area and web thickness for an assumed span to depth ratio. This will vary from about 12 for floor loading to 18 for roof loading (no access). If the beams are widely spaced the above ratios may not be appropriate. Before considering the detailed design of plyweb beams it will be necessary to survey methods of evaluating the strength properties of plywood.

PLYWOOD—STRENGTH PROPERTIES

A sheet or panel form of wood can be obtained by rotary cutting of logs to produce thin veneers (plies) which are glued together under pressure with the grain direction of alternate layers at right angles. Plywood has all the structural advantages of wood and at the same time eliminates some of its inherent disadvantages.

As the alternate layers are glued together at right angles cross grain strength is automatically achieved and moisture movement, which is predominantly across the grain, is restrained by the crossed layers. The strength and durability of plywood has long been established in the aircraft and building industries (see Chapter 3) and methods of evaluating its strength properties are outlined below.

Parallel plies approach

Large scale structural applications of plywood originated in North America (Douglas Fir Plywood) and the design approach in tension, compression and flexure is to ignore the plies which are stressed at right angles to the grain. The resistance of timber to stress at right angles to the grain is low Fig. 9.6 and thus the error involved in making this assumption should be small. However in the case of a three-ply laminate subjected to bending stress with its face grain direction perpendicular to the span, the face plies add considerably to its stiffness

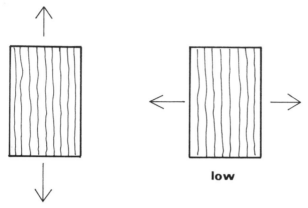

low

Fig.9.6 The resistance of timber to stress at right angles to the grain is low.

and a modification factor is introduced in the standard bending equation as below (18)

$$M = K.f.Z \qquad\qquad 9(ix)$$

where M is the applied bending moment
   f the maximum working stress for the extreme fibre
   Z the section modulus calculated on the basis of parallel or perpendicular flies only
and K = 1.5 for 3-ply plywood having its face grain perpendicular to the span
   K = 1.0 for plywood of 5 or more plies having its face grain perpendicular to the span
   K = 0.85 for all other cases.

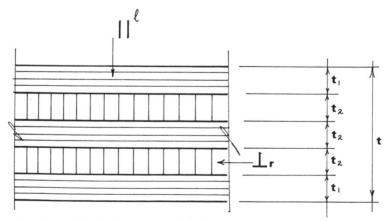

Fig.9.7 Ply thicknesses with face grain parallel to the span (t = 12.0 mm).

Typical calculations for the evaluation of the section properties of 5-ply, Douglas Fir plywood are given below. Consider ply thicknesses as in Fig. 9.7 with face grain in a direction parallel to the span and a 1 m width. $t = 12.0$ mm, $t_1 = 1.5$ mm, $t_2 = 3$ mm

$$2^{\text{ND}} \text{ Moment of area I} = \frac{1000}{12} [t^3 - (3t_2)^3 + t_2{}^3] \qquad\qquad 9(x)$$

$$= \frac{1000}{12} [12^3 - (3 \times 3)^3 + 3^3]$$

$$= 85.8 \times 10^4 \text{mm}^4/\text{m}$$

Section modulus Z $\quad = \dfrac{2I}{t}$

$$= \frac{85.5 \times 10^4 \times 2}{12} = 14.25 \times 10^3 \text{ mm}^3/\text{m}$$

Effective thickness of parallel plies

$$= 2 \times 1.5 + 3$$

$$= 6.0 \text{ mm}$$

Effective area of parallel plies

$$= 6.0 \times 1000$$

$$= 6.0 \times 10^3 \text{ mm}^2$$

Shear is often a critical force action in the design of plywood members due to the occurrence, in theory, of rolling shear (see Chapter 5). Shear in the plane of the plies produces a tendency for the wood fibres in the ply at right angles to the shearing force to roll. Further data on rolling shear is given in the section on ply-web box beams. Allowable stresses and design fundamentals for Douglas Fir plywood are available from Plywood Manufacturers of British Columbia (PMBC) (18), a division of the CFI and also in the commentary on CP112: 1971 (4).

Full cross section approach

Calculations for structural members based on the parallel plies only approach can be tedious and involve a number of modification factors. This has led to the development of the 'full cross section' approach which has been adopted by the Finnish and English Plywood manufacturers. From the results of tests on small specimens and built-up components the ultimate stresses obtained (4) were based on a full cross section, from which grade stresses can be derived using a formula of the same type as 9(i).

Grade stresses for Finnish European Birch Plywood (Finply—Exterior) are given in CP112 and further data is available from the Finnish Plywood Development Association. Supplies of solid birch plywood have become less readily available in the United Kingdom and an alternative is to use two species of plywood having birch exterior plies and a softwood core (spruce).

The strength properties of the spruce cored material are somewhat less than those for the solid birch plywood and data is available from F.P.D.A., again based on the full cross section approach.

It should be noted that in CP112: 1971 the stresses for Canadian Douglas Fir plywood have been modified so that they can be used with the 'full cross section' to maintain a consistent approach with that for European Redwood. Examples of calculations for timber components incorporating plywood are given in subsequent sections of this chapter.

Ply web beams—detailed design

Preliminary dimensions for the chords and webs may be obtained from equations 9(vii) and 9(viii) and a detailed section analysis is carried out as follows. If the modulus of elasticity of the chord material is significantly different from that of the web the plywood may be replaced by an equivalent thickness of the timber used for the chords. Thus the equivalent thickness will be given by t = ply thickness × E ply/E chord. The value of E ply is not the value given for bending as this is relevant to loading perpendicular to the plane of the plywood. It is appropriate to use the modulus of elasticity in tension and compression for the face grain parallel or perpendicular to the span. From a manufacturing point of view it can be more economic to use plywood with the face grain perpendicular to the span. It is also recommended that for the chords the mean elastic modulus value is used. The mean value is suggested by TRADA (19) and the authors would confirm this from the results of load tests (see Chapter 15). The following reasons for adopting the mean value are listed by TRADA:

(1) There are at least two flange members to average out their properties to some extent
(2) A high standard of construction using seasoned timber is essential for glued beams

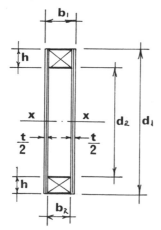

Fig. 9.8 Section properties of a ply web beam.

(3) Extensive experience in practice and the results of load testing have justified the use of the mean value, and
(4) Using the mean value leads to grossly erroneous estimates of the way in which the moment is shared between the chords and webs.

If the beam consists of Finply webs and species group S2 chords then E chord ~E ply a homogeneous section may reasonably be assumed. From Fig. 9.8 the section properties are

$$\text{Cross sectional area A} = b_1 d_1 - b_2 d_2 \qquad 9(\text{xi})$$

$$\text{Second moment of area } I_{xx} = \frac{1}{12}\left[b_1 d_1{}^3 - b_2 d_2{}^3\right] \qquad 9(\text{xii})$$

$$\text{Section modulus Z} = \frac{1}{6d_1}\left[b_1 d_1{}^3 - b_2 d_2{}^3\right] \qquad 9(\text{xiii})$$

Bending

The maximum bending stress f_b is given by

$$f_b = \frac{M}{Z}$$

Panel shear

The maximum panel shear stress will occur at the mid depth of the section and is given by

$$q_p = \frac{QA\overline{Y}}{It} \qquad 9(\text{xiv})$$

where   Q is the applied shear force
$A\overline{Y}$ is the first moment of area of the material above the neutral axis line x — x about line x — x thus

$$A\overline{Y} = b_2 h\left(\frac{d_1}{2} - \frac{h}{2}\right) + \frac{d_1{}^2}{8} t \qquad 9(\text{xv})$$

$$= \frac{b_2 h}{2}(d_1 - h) + \frac{t d_1{}^2}{8} \qquad 9(\text{xv})$$

Rolling shear

A critical factor in the design of ply-web beams is the transfer of shear from the chords to the webs. At the chord web interface shear will occur in the plane of the plies and thus the wood fibres will tend to roll. It is normal practice to glue the webs to the chords and rolling shear is a type of failure which can occur with this type of joint. If the face grain is parallel to the span then the maximum rolling shear stress will occur in the second ply from the chord, and where the face grain is perpendicular to the span it will occur in the ply glued to the chord, Figs. 9.9 and 9.10.

The rolling shear stress is calculated from the following expressions (Fig. 9.8)

$$q_R = \frac{QA\overline{Y}}{2Ih} \qquad 9(\text{xvi})$$

where $A\overline{Y} = \dfrac{b_2 h}{2}(d_1 - h)$

For rolling shear at the junction of the web and chord of a plyweb beam CP112: 1971 limits the maximum stress to 50 per cent of the grade stress.

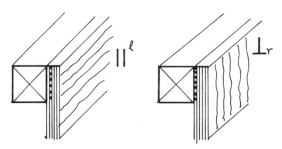

Figs.9.9 & 10 Rolling shear stress in glued joint.

Deflection

The deflection of the beam due to bending may be calculated in the usual way. If E chord differs considerably from E ply then it will be necessary to convert the ply to an equivalent thickness of the material used for the chords and thus the value of I will be that for the equivalent beam consisting of chord material only. For uniformly distributed load the deflection due to bending is given by

$$\delta b = \frac{5}{384}\frac{WL^4}{EI} \qquad\qquad 9(xvii)$$

In the design of ply web beams equation 9(xvii) may underestimate the total deflection as plane cross sections do not remain plane in beinding. The shear deflection δq should be added to the bending deflection to give the total deflection δt

$$\delta t = \delta b + \delta q \qquad\qquad 9(xviii)$$

The shear deflection may be calculated using the principle of virtual work. (Fig. 9.11)

The shear deflection is given by

$$\delta q = F \int \frac{Q_1 Q_0 ds}{GA}$$

where F is a form factor dependent on the cross sectional dimensions of the beam. Timoshenko and Gere give the following values (20).

Section	F form factor
Rectangle	6/5
I or box section	$\dfrac{\text{Area of cross section}}{\text{area of web}}$

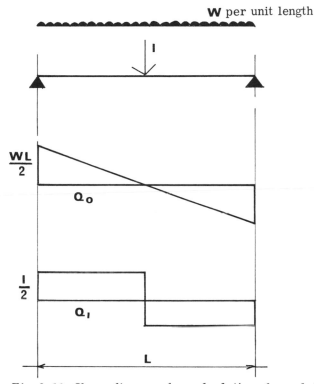

W per unit length

Fig.9.11 *Shear diagram for calculating shear deflection*

Q_0 is the applied shear force, Q_1 the shear force due to unit load applied at the point at which the deflection is required, G is the modulus of rigidity and A the cross sectional area.

$$\delta q = \frac{F}{GA}\frac{L}{2}\frac{1}{6} \cdot 2\left[\frac{WL}{2}\cdot\frac{1}{2} + 4.\frac{WL}{4}\cdot\frac{1}{2} + 0\right]$$

$$= \frac{FL}{GA6}\left[\frac{WL}{4} + \frac{WL}{2}\right]$$

$$= \frac{3}{4}\frac{WL^2F}{6GA}$$

$$= \frac{WL^2F}{8GA} \tag{9(xix)}$$

Thus the total deflection δt is given by

$$\delta t = \delta b + \delta q = \frac{5}{384} \frac{WL^4}{EI} + \frac{WL^2 F}{8GA}$$

9(xx)

$$= \frac{5WL^4}{384EI} \left[1 + \frac{48FEI}{5GAL^2} \right]$$

A similar procedure can be adopted for other loading conditions. If the depth to span ratio of the beam is less than $\frac{1}{12}$ an approximation to the total deflection (δt = δb + δq) is obtained by multiplying the bending deflection δb by a factor of 1.15.

The length of ply web beams is such that joints are required both in the chords and the webs. Loading tests on ply web beams carried out by the author have shown that in general failure occurs at a joint and this is considered further in a subsequent section of this chapter.

It is common practice to insert vertical stiffeners at intervals of about twice the beam depth to prevent local buckling of the ply webs and consideration should also be given to the possibility of lateral buckling of the compression flange. Useful guidance on this is given by CFI (21) and the compression flange is assumed to act as a column between points of lateral support.

Finally it is possible to camber ply web beams during manufacture by means of jigs. For roof construction it is a relatively easy matter to camber out the dead load deflection.

LAMINATED BEAMS

The laminated beam differs in construction from plywood insomuch that all the boards are assembled with the grain approximately parallel. Horizontally laminated members have the boards running parallel to the span of the member and vertically laminated members have the boards running at right angles to the span (Figs. 9.12 and 13). The

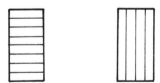

Figs. 9.12 & 13 Horizontally laminated members have boards parallel to span, while vertically laminated ones are at right angles to the span.

laminations are bonded by an adhesive and the construction of laminated beams requires expensive equipment and considerable expertise.

Production costs are obviously higher than for sawn timber sections but there are a number of advantages:

(a) The use of adhesives and automated finger and scarf jointing techniques permits a more economical use of timber as the thin laminates (usual thickness up to 50 mm) are connected to form much larger lengths.

(b) It is possible to produce much larger sections than those available in sawn timber and also the depth can be varied to suit strength requirements.

(c) As the laminates are relatively thin they can more easily be kiln dried to a specified moisture content (12 to 15 per cent) than large sawn sections.

(d) The effect of defects in an individual board is much less significant when considering the defects in relation to the beam as a whole consisting of a large number of laminations.

(e) The large cross sectional area of laminated beams enables adequate fire ratings to be achieved without protective treatment. This is illustrated by means of a numerical example in Chapter 11.

From the above is is apparent that higher design stresses could be used with laminated beams than for solid timber of the same species. In CP112: 1971 laminating boards are classified into three grades:

LA.. equivalent to 75 grade
LB.. equivalent to 65 grade
LC.. equivalent to 50 grade

The grade stress for a horizontally laminated member is taken as the product of the basic stress for the timber and an appropriate modification factor relating to the grade of timber, number of laminates and type of force action. Further modification factors are given for members laminated from more than one grade of timber and for vertically laminated beams.

Consider a sawn timber section of European Whitewood, 50 grade. The associated dry stress for bending, ignoring the deep beam effect, is 8.75 N/mm². For a laminated beam of the same species, LB grade, the basic stress for beinding is 14.7 N/mm² and the appropriate modification factor (see table 5 CP112) for say 15 laminations is 0.8. Thus the grade stress is $14.7 \times 0.8 = 11.75$ N/mm². This value is 34 per cent higher than that for the 50 grade sawn timber. Another important comparison between sawn-timber and laminated beams is that for the latter the mean modulus of elasticity can be used. Typical cross sectional dimensions (b × d) for horizontally laminated timber beams range from 90 mm × 300 mm to 185 mm × 900 mm.

STRESSED SKIN PANELS

Solid timber framing members covered with a top and/or bottom skin of plywood form stiff panels which are widely used for floors, roof and walls (Fig. 9.14). Interaction between the solid timber joists and the

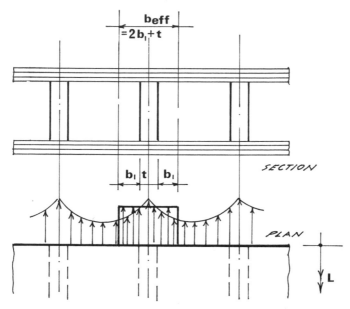

Fig. 9.14   Theoretical analysis of stressed skin panel.

plywood is achieved by a glued connection and structurally the panel
may be considered to act as a series of T or I beams. With a properly
glued plywood-timber joint it may be assumed that there is no slip
at the interface. The term stressed skin is used to define the structural action of the panel as the assembly behaves as a series of linked I
or T beams the ribs of which deflect under the applied loading and
direct forces are induced in the plane of the skin. A theoretical analysis of stressed skin panels is somewhat complex as the direct forces
are not uniformly distributed across the width of the skin (Fig. 9.14).
For design purposes it is convenient to consider a width of the skin,
termed the effective breadth, acting with the rib, which, if stressed
uniformly would give the same contribution to the flexural strength of
T or I beams as the whole of the non-uniformly stressed skin. Tottenham (22) has shown that the effective width of the skin on each side of
the joist can be determined as a proportion of the span. The effective
width b_1 on either side of the joist can be expressed in the form (Fig.
9.14).

$$b_1 = \frac{L}{K}$$

9(xxi)

Where L is the span in inches and k = 5.3 + 9.0n for Douglas Fir.
The factor n represents the proportion of plies with the grain running
in the direction in which the panel spans. Values of n should lie between 0.25 and 0.75.

An alternative design method has been published by the P.M.B.C. (23)
again relating to the use of Douglas Fir. The width of the skins is

governed by a basic clear distance b between logitudinal members obtained from the following formula

$$b = 31h\left[\frac{h}{e}\right]^{1/2} \text{ for 3-ply plywood} \qquad \text{9(xxii)}$$

$$b = 36h\left[\frac{h}{e}\right]^{1/2} \text{ for 5 or more ply-plywood} \qquad \text{9(xxiii)}$$

where h = thickness of plywood
       e = total thickness of plies parallel to the span.

If the clear distance between the longitudinal members exceeds b then the determination of the neutral axis and second moment of area of the I section is based on a skin width of b plus the joist thickness. If b is greater than the clear distance between the framing members then skin width is taken as the joist spacing. Only the plies in the skins which have their grain parallel to the span should be considered in the above calculations. Working stresses should be based on PBMC recommendations (23) and it should be noted that for flexure and compression parallel to the grain the allowable stresses are reduced from 100 per cent when the clear distance between joists is b/2 linearly down to 69 per cent when the clear spacing is b or greater. Tabulated data on the span range and load carrying capacity is readily available (23) which avoids the need to carry out the calculation procedures outlined above.

TRUSSED ASSEMBLIES

Apart from the solid floor joist light timber trusses are the most widely used timber structural component. There are a number of types available for commercial and particularly for domestic use and modern manufacturing techniques utilising patented connector plates enable the truss to be produced in large quantities with subsequent economic benefits. Typical of the domestic range available is the W-truss (Fig. 9.15) which is manufactured by a number of timber engineering organisations. They are designed to carry a total load in the order of 2.0 KN/m² (self weight + imposed loading). A common spacing is 0.6 m

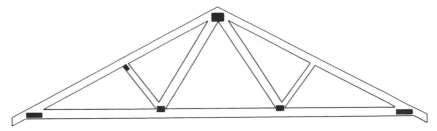

Fig.9.15 Typical W-truss showing positions of gang-nail plates that are forced into place by hydraulic pressure

with a pitch variation of $17\frac{1}{2}$ to 35° and spans up to about 10 m. The width of the timber framing members is seldom greater than 37.5 or 50 mm and the completed truss is a rigid structure, due to its triangulation, light and easily transported, stored and erected. Before the development of the 'Gang-Nail' connection plate the connections between the members tended to be cumbersome although the basic analysis as a pin jointed plane framework is straightforward. The Gang-Nail machine produces trusses which have all their members in one plane without projections, the plates being pressed into position hydraulically.

For non-standard spans and loadings the member forces may be obtained analysing the truss as a pin jointed framework using simple statics. The truss fabricator should be consulted with regard to economic member sizes and connection details. Timber trusses have little lateral stability and are vulnerable to the 'pack of cards' type of failure during erection. The erection procedure is to fix the trusses at the wall plates and to tie them together—a battern can be nailed near the apex and also at ceiling joist level.

PORTAL FRAMES

Portal frames of laminated or ply-box/I construction have found a wide range of applications—agricultural and industrial buildings to

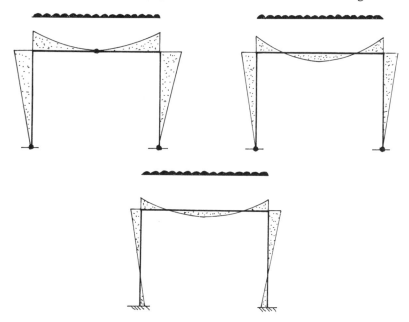

Fig. 9.16 Portals: pinned at feet and crown, two pinned, and fixed.

swimming pools, assembly halls and churches. At the initial design stage attention should be given to proportioning the frame to give a minimum amount of material compatible with functional, transportation and erection considerations. It is generally more convenient from the analytical and constructional standpoint to design the frame with the assumption that it is pinned at the feet and crown. However a more even moment distribution is obtained with a two pinned or fixed portal Fig. 9.16.

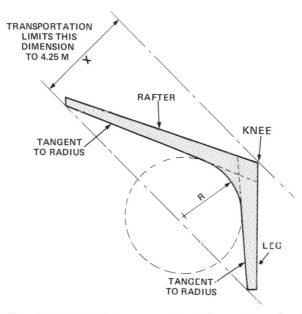

TRANSPORTATION
LIMITS THIS
DIMENSION
TO 4.25 M

RAFTER

KNEE

TANGENT
TO RADIUS

R

LEG

TANGENT
TO RADIUS

Fig. 9.17 Portal frame critical dimensions for transport

In general the rafter portion should not be flat as this increases the knee moment and the difficulty of transport. The critical dimension for transport is x, Fig. 9.17, and this is limited to about 4.25 m. Consider a portal frame with a flat and pitched rafter (Figs. 9.18 and 19). In the case of the flat rafter moments can be taken about the crown hinge to find the horizontal reaction, notation as in diagram

$$\frac{W}{2} \cdot \frac{L}{2} - H.h - \frac{W}{2} \cdot \frac{L}{4} = 0$$

$$H = \frac{WL}{8h} \qquad\qquad 9(\text{xxiv})$$

Bending moment at knee $= \frac{WL}{8h} \cdot h$

$$= \frac{WL}{8} \qquad\qquad 9(\text{xxv})$$

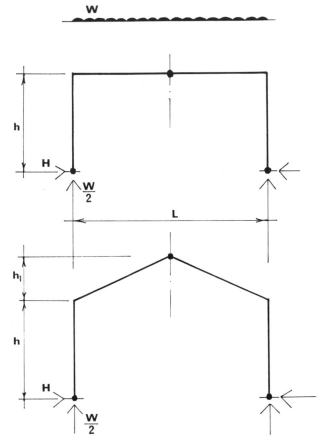

Figs. 9.18 & 19   Portal frames with flat and pitched rafter.

For the pitched portal taking moments about the crown

$$\frac{W}{2}.\frac{L}{2} - H(h + h_1) - \frac{W}{2}.\frac{L}{4} = 0 \qquad\qquad 9(\text{xxvi})$$

$$H = \frac{WL}{8\,(h + h_1)} \qquad\qquad 9(\text{xxvii})$$

$$\text{Bending moment at knee} = \frac{WL}{8}.\frac{h}{(h + h_1)} \qquad\qquad 9(\text{xxviii})$$

The knee moment is less for the pitched portal and in the case of laminated portals there are some limitations imposed on the radius at the knee and the pitched rafter thus has a further advantage. In the fabrication of curved laminated members bending stresses are induced in the laminates and investigations have shown that these stresses are related to the lamination thickness t and the radius of curvature R. This initial bending stress reduces the strength of the member and thus for the curved portion of a laminated member the grade bending stress should be multiplied by a modification factor.

The following values are given in CP112: 1971

t/R	$\frac{1}{100}$	$\frac{1}{125}$	$\frac{1}{150}$	$\frac{1}{175}$	$\frac{1}{200}$	$\frac{1}{250}$	$\frac{1}{300}$
Modification factor to be applied to grade bending stress	0.8	0.87	0.90	0.91	0.93	0.94	0.95

It is recommended that t/R should exceed $\frac{1}{125}$ for softwoods and $\frac{1}{100}$ for hardwoods. Thus for a 25 mm thick laminate the minimum radius is 3.125 m. The section analysis of portal frames involves consideration of bending, shear and axial forces. The combination of bending and axial forces is covered in the next section.

COLUMNS

The design of timber columns follows the usual pattern for compression members in which account is taken of slenderness, end restraint and initial curvature. As with other structural materials a starting point for the design of columns is the Euler formula in which the critical load P is given by

$$Pcr = \frac{\pi^2 EI}{l^2} \qquad 9(xxix)$$

where   l = pin ended strut end
   E = modulus of elasticity of the material
   I = second moment of area of section about its weaker axis

Putting I = Ar² where A is the cross sectional area and r the radius of gyration equation 9(xxix) may be written as

$$Pcr = \frac{\pi^2 EA}{\left[\frac{l}{r}\right]^2} \qquad 9(xxx)$$

where l/r is known as the slenderness ratio.

The assumptions made in the development of the Euler formula are never realised in practice and the Perry Robertson formula which is discussed in standard text books on the strength of materials embraces the ultimate compressive strength of the material and initial curvature. For values of l/r in excess of about 80 the Euler and Perry Robertson formulae are in close agreement as the modulus of elasticity of timber is assumed to remain constant in a particular species irrespective of the stress grade. Inspection of equation 9(xxx) shows that no material economy is gained by using the higher grade material. The Perry Robertson formula is cumbersome to apply to timber columns as it takes into account the ultimate compressive strength parallel to the grain for a particular grade of timber, the Euler stress $\pi^2 E / \left[\frac{1}{r}\right]^2$ and an eccentricity coefficient. Booth and Reece consider the problem in some detail in the commentary on CP112 (4) and this document should be consulted for an explanation of the modification factors to be applied to the grade stress in compression parallel to the grain to give a permissible compressive stress for a given slenderness ratio.

CP112: 1971 gives the following effective length values for determining the slenderness ratio of compression members where L is the actual length of the member

Condition of end restraint	Effective length l
Restrained at both ends in position and direction	0.7L
Restrained at both ends in position and at one end in direct	0.85L
Restrained at both ends in position but not in direction	L
Restrained at one end in position and direction and at the other end partial restraint in direction but not in position	1.5L
Restrained at one end in position and direction, but not restrained in position and direction at the other end	2.0L

The code gives the following modification factors for slenderness ratio and duration of loading on compression members of 40 grade and 50 grade softwoods.

TABLE 9.2   SLENDERNESS RATIO l/r

r = minimum radius of gyration	Modification factor		
	long-term loads	medium-term loads	short-term loads
less than 5	1.00	1.25	1.50
5	0.99	1.24	1.49
10	0.98	1.23	1.47
20	0.96	1.20	1.44
30	0.94	1.17	1.40
40	0.91	1.13	1.34

TABLE 9.2   SLENDERNESS RATIO l/r (continued)

r = minimum radius of gyration	Modification factor		
	long-term loads	medium-term loads	short-term loads
less than 50	0.87	1.08	1.27
60	0.83	1.00	1.16
70	0.77	0.90	1.01
80	0.70	0.79	0.86
90	0.61	0.68	0.72
100	0.53	0.58	0.60
120	0.40	0.42	0.44
140	0.31	0.32	0.33
160	0.24	0.25	0.25
180	0.20	0.20	0.20
200	0.16	0.16	0.17
220	0.13	0.14	0.14
240	0.11	0.12	0.12
250	0.10	0.11	0.11

The above factors are not safe for higher grade softwoods and all hardwoods and the code gives another set of modification factors.

As an illustrative example of the application of the above modification factors consider a timber strut of length 2.4 m, cross sectional dimensions 100 mm × 50 mm (d × b) restrained at both ends in position only. Determine the permissible stress in compression parallel to the grain for short term loads if the grade stress is 6 N/mm^2

$$r = \left[\frac{I}{A}\right]^{1/2}$$

$$= \left[\frac{db^3}{12db}\right]^{1/2} \quad \text{where } d > b$$

$$= \frac{b}{\sqrt{12}} = \frac{1}{3.47} b$$

thus $r = \frac{1}{3.47} \times 50$

$= 14.4$ mm

$1 = 2400$ mm

Slenderness ratio $\dfrac{l}{r} = \dfrac{2400}{14.4}$

$$= 167$$

From Table 9.2 the modification factor approximates to 0.23 for short term loads and thus the permissible stress = 0.23 × 6 = 1.38 N/mm^2, less than one quarter of the grade stress.

The code imposes a limiting value of 180 for the slenderness ratio for members carrying loads resulting from dead and imposed loads. This may be increased to 250 for members carrying loads resulting from wind loads only.

In many practical designs the combination of compression and bending must be considered and CP112 gives the usual interaction formula in terms of stress ratios that is

$$\left.\dfrac{f^c \text{ applied}}{f^c \text{ permissible}} + \dfrac{f^b \text{ applied}}{f^b \text{ permissible}} \not> \right] \begin{array}{l} 1.0 \text{ for } l/r \not> 20 \\ 0.9 \text{ for } l/r > 20 \end{array}$$

where f^c applied   =   applied compressive stress parallel to the grain

        f^c permissible =   permissible compressive stress parallel to the grain

        f^b applied   =   applied bending stress parallel to the grain

        f^b permissible =   permissible bending stress parallel to the grain

FOLDED PLATE PANELS

The maximum span of the flat stressed skin panel referred to earlier in the chapter rarely exceeds a few metres and practical depths are up to about 0.15 m. The folded panel is an ingeneous adaptation of stressed

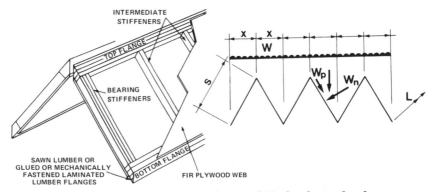

Fig.9.20 Elements of a folded plate and its load resolved into components.

skin construction to form large span roofs of saw-tooth cross section generally referred to as plywood folded plates. The basic components of the folded plate are shown in Fig. 9. 20 and as with the concrete folded plate an understanding of its structural action can be obtained by resolving the load into components parallel and perpendicular to the plane of the plate (Fig. 9. 20). For the perpendicular component of the load the effective structural depth is t and the structure is designed as a stressed skin panel spanning a distance s from the ridge to the valley. The effective structural depth for the component of the load parallel to the plane of the plate is s and the structure is designed as a plywood web beam of span L, the chord material being at the ridge and valley.

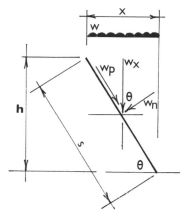

Fig.9.21 *A single plate of multifold system analysed in terms of simple statics.*

A single plate of multifold system may be analysed in terms of simple statics. Assuming the applied loading w per unit of length to be uniformly distributed in plan the total load wp in the plane of a single plate (Fig. 9. 21) is given by:

$$wp \sin\theta = wx$$

$$wp = \frac{wx}{\sin\theta}$$

9(xxxi)

The total load carried perpendicular to the plane of the plate is

$$= wx \cos\theta$$

$$= w.x.\frac{x}{s}$$

Thus the perpendicular load per unit area is given by

$$= w.\frac{x^2}{s^2}$$

$$= w \cos^2 \theta$$

9(xxxii)

the horizontal component of the load wp is given by

$$wh = wp \cos\theta$$

$$= wx \cdot \frac{\cos\theta}{\sin\theta}$$

thus $wh = wx \cos\theta$

$$= wx \frac{x}{h}$$

$$= \frac{wx^2}{h} \qquad\qquad 9(xxxiii)$$

The structure is designed as a beam of depth S, span L for the loading wp. Thus, assuming the distance between the centroids of the chords to be 0.95S an approximation to the chord area A required is given by:

$$A = \frac{Mp}{fp \times 0.95} \qquad\qquad 9(xxxiv)$$

where Mp = bending moment due to $wp = \frac{wpL^2}{8}$

and fp = permissible stress in chord material

The maximum shear $Qp = \frac{wpL}{2}$ $\qquad\qquad 9(xxxv)$

The design procedure thus takes the same form as the plywood web beams considered earlier in the chapter. The structure is treated as a stressed skin panel of span S for the component Wn giving

$$Mn = \frac{Wns^2}{8} \qquad\qquad 9(xxxvi)$$

$$Qn = \frac{Wns}{2} \qquad\qquad 9(xxxvii)$$

Considering the horizontal component of the loading wh, the total thrust at the end supports will be given by

$$H = \frac{WhL}{2}$$

$$= \frac{Wx^2}{h} \cdot L \qquad\qquad 9(xxxviii)$$

For the interior valley junctions (Fig. 9.22) the thrusts will be balanced under symmetrical loading only, but there will be an unbalanced force at the end plate. Unbalanced thrusts can be resisted by ties or diaphragms and the diaphragms will also help to maintain the geometry of the folded system. Comprehensive design data and typical details for plywood folded plates has been published by PMBC (24).

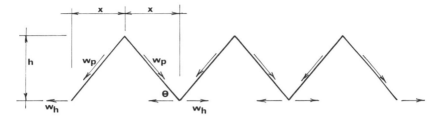

Fig. 9.22 Interior valley junctions: horizontal thrusts balanced under symmetrical loading

Correct structural detailing of plywood folded plates is essential and reference should be made to the PMBC publication if a detailed design is to be carried out.

CONNECTIONS

The design of timber connections involving the transfer of force actions from one member to another has been the subject of much ingenuity. In the design of large span roof trusses Renaissance engineers developed spliced joints to transmit tension and a typical detail is illustrated in Fig. 2.8 page 19. More recently a patent was issued (25) for machine-dadoed, sawtooth-notched, interlocking and overlapping joints which can transmit either tensile or compressive forces (Fig. 9.23).

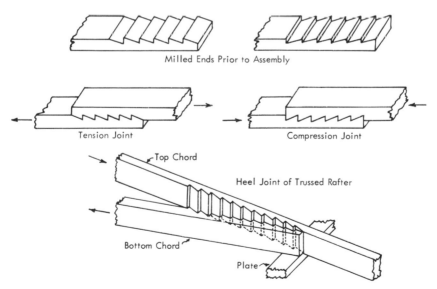

Fig. 9.23 The Fellman joint: a machine dadoed, sawtooth notched, interlocking and overlapping timber joint.

Early designs for timber joints had several drawbacks—overlapping members were in two planes, causing load eccentricity, the total assembly thickness was large and timber shrinkage could cause excessive distortion. Recent advances in timber engineering, particularly in the automated production of timber components, assembled on special jigs has to a large extent eliminated the need for machined splice joints in trusses and it is now general practice to use metals in the form of nails, screws, bolts and connector plates to form connections between timber members. Due to the number of variables involved, a theoretical analysis of timber connections incorporating metal fasteners is not in general practical, and working loads are normally based on test results. The working load can be obtained from the formula used for obtaining basic stresses for timber specimens, that is

$$\text{working load} = \frac{P - k\sigma}{\text{F.S.}} \qquad\qquad 9(\text{xxxix})$$

where P = average ultimate load obtained from test results
        σ = standard deviation
        k = 2.33 for one per cent exclusion, see Chapter 5
   F.S. = factor of safety

The value used for the factor of safety is a matter of engineering judgement but it should be of such a value that the working load is less than that which exceeds the limit of proportionality, or causes excessive slip. It should be noted that the load-slip relationship for timber joints

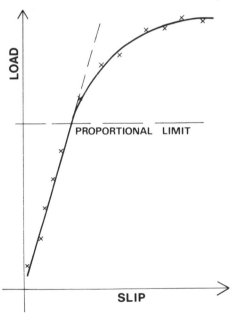

Fig. 9.24 Load slip relationship for timber joints incorporating metal fasteners.

incorporating metal fasteners such as nails, screws and bolts takes the form shown in Fig. 9.24 and even under short term loading it is necessary to ensure that permanent joint deformation does not occur. Initial slip at the early stages of loading cannot be avoided as movement will take place until the metal is bearing firmly against the timber.

The commentary on CP112 offers guidance for the design of joints incorporating nails, screws, bolts, toothed and split ring connectors and glues. The scope of this information is not sufficient to cover the large selection of fasteners available as these are in a state of continuous development. Thus the designer must rely on manufacturers data which should be based on equation 9(xxxix).

It is not possible in this introductory text to cover all recent developments and this section will be limited to a brief survey of the data given in the commentary on CP112, fastenings in plywood and Gang-nail connector plates. The strength of a joint incorporating nails, screws, bolts or connectors is related to the species of timber used and for design purposes softwoods and hardwoods are classified into four groups J1, J2, J3 and J4 (table 21 CP112). The loads recommended in CP112 for nails, screws, bolts and connectors are applicable to long term loading and where duration of load is being considered the following modification K_J applies.

$$K_J = \frac{1 + K_D}{2} \text{ (nails screws and tooth plate connectors)}$$

$$K_J = K_D \qquad \text{(bolts, split rings and shear plates)}$$

where K_D is a duration of load factor defined in Chapter 8.

NAILED JOINTS

The code gives basic loads for round wire nails loaded laterally or in withdrawal (CP112 tables 22 and 23 respectively). There are many types of nail available which give an improved performance and the code recognises this by allowing a 25 per cent increase in basic lateral loads for square grooved or square twisted nails of steel with a yield stress of not less than 385 N/mm^2. The nominal diameter of the nail should be assumed to be 0.75 times the distance between diagonally opposite corners of the cross section. The strength of a nailed joint under lateral load is dependent on a number of factors, the thickness of the timber members, the bending strength of the nails and the compression strength of the timber, end and edge distances and the moisture content of the timber. In table 22, referred to above, the basic lateral loads (dry timber) are related to the nail diameter, standard thicknesses of members headside and pointside Fig. 9.25 and the species group (J1 to J4). These loads are based on tests carried out at the Forest Products Research Laboratory, Princes Risborough. The ultimate load P_L for a two member joint, nails in single shear is given by the expression (4)

$$P_L = 262 \, Gd^2 \, N \qquad\qquad\qquad 9(xl)$$

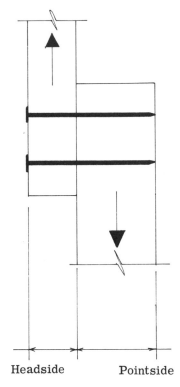

Headside   Pointside

Fig. 9.25 Headside and pointside of a nailed joint.

where   d = nail diameter mm
            G = specific gravity based on oven dry weight and volume at test

The basic lateral load is derived from equation 9(xl) as follows. The value of G is obtained by subtracting 2.33 times the standard deviation from the mean value. The basic load is then evaluated using a load factor of 3.0. This load factor is higher than that used for obtaining basic stress for timber members (2.25) as it is considered that poor workmanship is more likely in the fabrication of nailed members. Consider a group J3 timber, Commercial Western Hemlock, for which the mean specific gravity is 0.41 and standard deviation 0.069 (3). For one per cent exclusion the characteristic specific gravity is 0.41 − 2.33 × 0.069 = 0.26. Thus from equation 9(xl) the basic lateral load for a 4 mm diameter round wire nail is

$$= \frac{262}{3} \times 0.26 \times 4^2$$

$$= 363 \text{ N}$$

This figure agrees reasonably well with the figure of 400 N for J3 timber given in CP112 table 22.

Basic resistance to withdrawal of round wire nails inserted at right angles to the grain is obtained from CP112 table 23 and the values are derived from the following expression for average ultimate load

$$P_W = 47.6 \; G^{5/2} d \qquad\qquad 9(\text{xli})$$

where P_W = ultimate load per nail in Newtons per mm of penetration

   G = specific gravity based on weight and volume when oven dry

   d = diameter of nail mm

The values of G are taken as follows

Group	G
J1	0.6
J2	0.51
J3	0.42
J4	0.34

A load factor of 6 is used to determine the basic loads which embraces duration of load and variability of material. The value of P_W should be divided by 24 in cases where nails are subjected to cycles of wetting and drying. The code also gives guidance on nail spacing and allows a 25 per cent increase in load where a metal component or plate of adequate strength is nailed to a timber member.

Equations 9(xl) and (xli) do not apply to fastenings in plywood for which there is little data available. From a limited number of tests carried out at the University of Surrey using 50 mm square twisted nails of 3.1 mm nominal diameter—assumed to be 0.75 times the distance between diagonally opposite corners of the cross section, driven through 12 mm birch ply into 100 mm × 75 mm Douglas Fir, failing loads (lateral) of 1.5 KN per nail were obtained. Using a load factor of 3.0 the basic load is 0.5 KN. Janson (26) suggests allowable loads for nailed joints with plywood (Swedish) and gives recommendations for distances between nails.

SCREWS

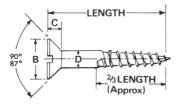

Fig.9.26 Standard dimensions of screw.

As with nails, tables for lateral and withdrawal loads are given in CP112 tables 25 and 26 based on test data. The lateral load carrying capacity of a screw depends on the species of the timber (J1 to J4), the diameter of the screw and the depth of penetration into the point side member. If the standard penetration is defined to be seven times the shank diameter (Fig. 9.26) of the screw then the lateral load is given by the formula (4)

$$P = Kd^2$$

where P = load in Newtons
       d = diameter of the screw, mm
       K = a constant depending on the species

The basic load is obtained by dividing the proportional limit by 1.6, giving a basic load in the order of one-sixth the ultimate. The following values of K apply to timber groups J1 to J4. For green timber a reduction factor of 0.7 is applied.

Group	K	
	moisture content $\leqslant$ 18 per cent	moisture content $>$ 18 per cent
J1	27.56	19.29
J2	22.74	15.92
J3	18.60	13.02
J4	14.47	10.13

Equation 9(xlii) and the value of K listed above applied to wood screws inserted at right angles to the grain.

The average ultimate load in withdrawal of common wood screws inserted into the side grain of dry wood is obtained from the expression (4)

$$P = 98.04\,G^2d \qquad\qquad\qquad 9(\text{xliii})$$

where P = ultimate load per screw in Newtons per mm of penetration of the thread
       G = specific gravity based on weight and volume when oven dry
       d = diameter of screw, mm

A load factor of six is applied to formula 9(xliii) and the basic loads for groups J1 to J4 are listed below. For green timber a reduction factor of 0.7 is applied.

Group	G	Basic Load Newtons	
		moisture content ⩽18 per cent	moisture content >18 per cent
J1	0.6	35.3 d	21.71d
J2	0.51	25.5d	17.85d
J3	0.42	17.3d	12.11d
J4	0.34	11.33d	7.93d

The above data applies to screws manufactured from steel meeting the provisions of BS 1210—wood screws.

BOLTS

CP112 table 27 gives dry basic loads on a bolt in a two member joint (single shear) which are related to:

(a)   the bolt diameter
(b)   thickness of thinner member

(c)   timber group (J1 to J4)
(d)   direction of load, perpendicular or parallel to the grain

As a starting point it would seem reasonable to design a bolted connection on the basis of multiplying the bearing area by the basic stress for compression perpendicular or parallel to the grain for the species of timber being considered. The distribution of stress set up under a bolt is however non uniform as illustrated in Fig. 9.27. Tests carried out by Trayer which are summarised in the code commentary (4) showed that the load slip relationship for three member connections (Fig. 9.28) is, apart from initial slip, linear up to a well defined proportional limit (Fig. 9.29). The magnitude of the average compressive stress parallel to the grain at the proportional limit was shown by Trayer to depend on the ratio L/D, where L is the length of the bolt in the centre

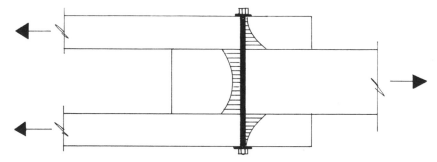

Fig. 9.27 Non uniform stress distribution under a bolt.

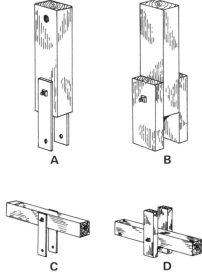

Fig.9.28 Bolted test specimens (top) to which load was
applied parallel to the grain: A, with metal side plates; B, with
wood side plates. (bottom) Bolted test specimens to which load
is applied perpendicular to the grain: C, with metal side
plates; D, with wood side plates.

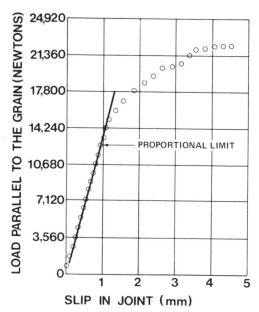

Fig.9.29 Graph of load slip relationship.

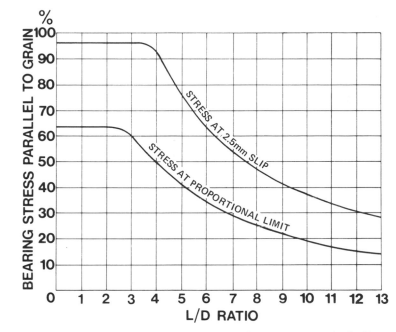

Fig. 9.30 For conifers, the relation between average bolt bearing stress and the ratio L/D.

member and D the bolt diameter. For conifers the relation between average bolt bearing stress (expressed as a percentage of the maximum compressive strength of the timber parallel to the grain) and the ratio L/D is shown in Fig. 9.30. These curves apply to joints with metal side plates. For the tests on joints with timber side plates (12.5 mm diameter bolts) each side piece was half the thickness of the centre member and the average proportional limit stress fell by 20 per cent. The L/D ratio is obtained using the thickness of the centre member. Consider a beam/column connection using a metal plate of thickness 5 mm, mild steel bolts of 12 mm diameter and baltic redwood (Fig. 9.31). An assessment of the capacity of this section will be made using Fig. 9.30. For baltic redwood, air dry, the average maximum compressive strength parallel to the grain is 45.0 N/mm^2 (1) with a standard deviation of 7.63 N/mm^2. Thus for one per cent exclusion the characteristic strength is $45 - 2.33 \times 7.63 = 27.22$ N/mm^2. The L/D ratio is $^{96}/_{12} = 8.0$, and from Fig. 9.30 the stress at 2.5 mm slip is $0.48 \times 29.22 = 13.06$ N/mm^2. Considering bolt line A-A the capacity of the connection based on 2.5 mm slip is

$$= 2 \times 13.06 \times 96 \times 12 \times 10^{-3} \text{ KN}$$

$$= 30.09 \text{ KN}$$

Finally it is necessary to assume an appropriate load factor. This

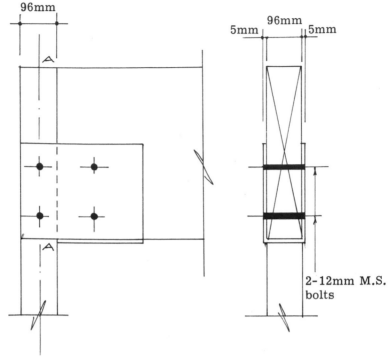

Fig.9.31 A beam column construction using a metal plate, mild steel bolts, and Baltic Redwood.

should embrace duration of loading and the possibility of poor workmanship. Using the factor of 3.0 as for nailed joints in lateral load the working load would approximate to 10 KN. The metalwork should also be checked to ensure it is not overstressed at this load. A further reduction factor of 0.7 should be applied to green and exposed timber. In cases where the load is applied at an angle to the grain it is necessary to take into account the considerable difference in the strength of timber in compression parallel and perpendicular to the grain. The formula due to Hankinson (4) is used to determine the basic load in a direction inclined to the grain:

$$N = \frac{PQ}{P \sin^2 \theta + Q \cos^2 \theta}$$

where θ = angle between the direction of the load and direction of the grain
P = load parallel to the grain, CP112 table 27
Q = load perpendicular to the grain, CP112 table 27

GANG-NAIL CONNECTOR PLATES

Punched metal plates enabling timber elements of uniform thickness to

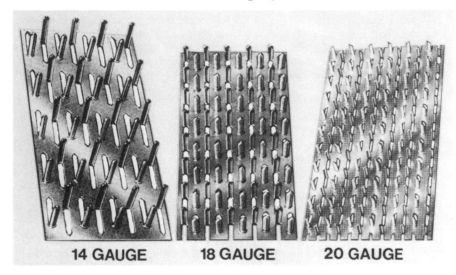

Fig.9.32 Gang-Nail punched metal plates.

be joined truss or panel configuration, were originally developed in the United States of America. The Gang-Nail punched metal plate, manu-factured by Automated Building Components (U.K.) Ltd., Farnham, Surrey. is extensively used in the United Kingdom for roof construction. The nails are stamped out from galvanised steel and have a curved profile with chisel ends Fig. 9. 32. The punched plates are pressed hydraulically into the timber (Fig. 9. 33) with minimum damage to the wood fibres. There are three gauges of Gang-Nail available (14, 18 and 20) with a large number of variations in size available for each gauge. This type of connector plate is not covered in CP112 but designs can be based on a report produced by the Forest Products Research Labor-atory describing strength tests on structural timber joints made with

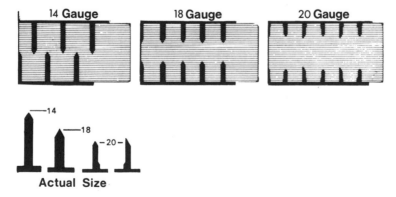

Fig. 9. 33 The profile of the nails in the three guages of the Gang-Nail plate.

20-gauge Gang-Nails (27). The FPRL report gives permissible loads per nail for a number of species of softwood. These loads relate to timber having a moisture content of not more than 22 percent at assembly and embrace the three categories of load duration. Gang-Nail connector plates are incorporated in a variety of truss types and wall panels manufactured by licenced fabricators in the United Kingdom.

SPLIT RING, SHEAR AND TOOTHED PLATE CONNECTORS

This type of connector has been widely used for industrial and domestic trussed assemblies. For domestic application it has largely been superseded by the Gang-Nail connector which is more suited to automated production techniques and avoids overlapping timbers at the connection, Fig. 9.34. However it has the advantage over a bolted connection that the load is transmitted to the timber in a more uniform manner (Fig. 9.35) thus reducing stress concentration (Fig. 9.27). The toothed plate or 'Bulldog' connector does not require special equipment and can be used for wood to wood or wood to metal joints. Split ring and shear plate connectors require special grooving tools as they are inserted into pre-cut recesses. Fig. 9.36 to 38 illustrate features of these connectors both of which are used for structures which are frequently dismounted. Basic loads and other data for connectored joints, toothed, split ring and shear plate are given in CP112 and the derivation of these loads from test data is summarised in the commentary (4). A design manual for the use of 'Bulldog'* round toothed plate connectors and 'Teco' split rings and shear plates is available from Macandrews and Forbes Ltd. (28).

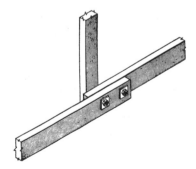

Fig. 9.34 Overlapping timber at the connection.

---

* This product is now marketed under the trade name 'MAFCO'.

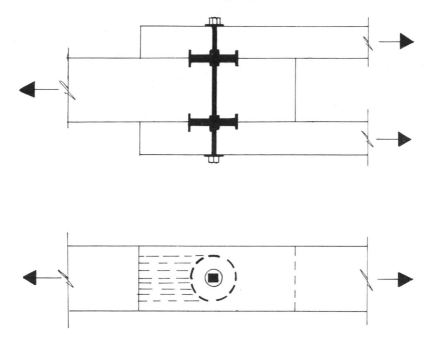

Fig.9.35 Wood to wood shear plate connector: gives more uniform load transmission over a greater area than with a bolted connection (see Fig.9.27)

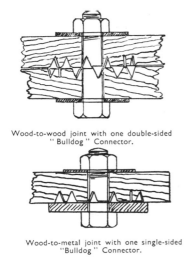

Wood-to-wood joint with one double-sided " Bulldog " Connector.

Wood-to-metal joint with one single-sided "Bulldog " Connector.

Fig.9.36 The toothed plate or 'Bulldog' connector.

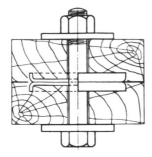

Wood-to-wood joint with two "Teco" Shear Plate Connectors.

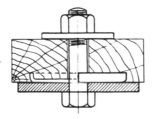

Wood-to-metal joint with one "Teco" Shear Plate Connector.

Fig. 9. 37 Arrangements for shear plate connectors.

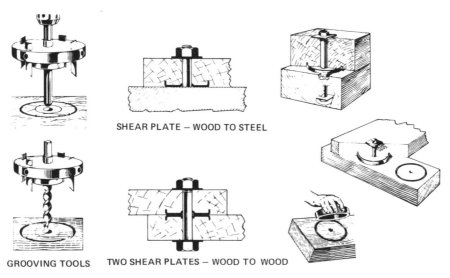

SHEAR PLATE – WOOD TO STEEL

GROOVING TOOLS   TWO SHEAR PLATES – WOOD TO WOOD

Fig. 9. 38 Detail of application of shear plate connectors.

GLUED JOINTS

Reliable synthetic adhesives have played a vital part in the development of timber engineering techniques in the past twenty years which are now geared to the automated production of structural components widely used in the domestic and industrial fields. The primary uses of adhesives are for the manufacture of plywood and in assembly work. The use of plywood and the range of adhesives available is surveyed in Chapter 3 and the following section is limited to the use of glues in the assembly of structural components such as box and I sections and load bearing panels.

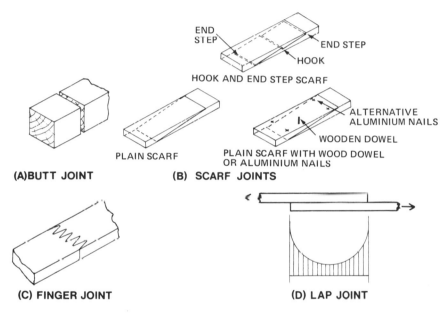

Fig. 9.39 Arrangements of butt, scarf, fingers, and lap joints.

A basic rule in the design of glued joints is that the joint should never be subject to direct tension alone in a direction at right angles to the plane of the contact surface. Considering some commonly used joints, Fig. 9.39, it can be seen that the butt joint should never be used to join lengths of timber subjected to tension. With the scarf and finger joints the stresses on the glue line may be resolved into shear and tension components and the steeper the slope of the joint the greater the tension stress at right angles to the glue line. The joint between the web and flange of box or I beams is stressed in shear (rolling) but, depending on the length of the chords, scarf or finger joints will be required where tension occurs. Limitations on the lengths of commercially available stocks of structural softwoods and hardwoods were mentioned earlier and thus end jointing will normally be required in box or I

beam chords with lengths in excess of about 6 m for European Redwood and Whitwood and 8 m for Douglas Fir and Hem Fir. Joints will
be required in the plywood webs at much closer centres, about 2.4 m
maximum.

It is commonly stated that the glue is stronger than the timber, where
the glue line is stressed in shear and thus permissible stresses for
glued joints will be those for timber in shear parallel or perpendicular
to the grain (rolling).

Values for European Redwood and Whitwood are listed below (dry),

TABLE 9.3

	Shear stress parallel to grain	Shear stress at right right angles to grain (rolling)
	N/mm^2	N/mm^2
Basic	1.52	0.38
75 grade	1.14	0.29
65 grade	0.97	0.24
50 grade	0.76	0.19
40 grade	0.62	0.16

The stresses given in Table 9.3 relate to glued joints of the type shown
in Fig. 9.39D but it should be noted that the actual stress distribution is
by no means uniform. In the use of a joint between a plywood web and
softwood chord the rolling shear values are appropriate. With regard
to scarf joints the strength of the joint is obtained by multiplying the
basic stress for the timber species by the efficiency rating appropriate
to the slope of the scarf, grade of timber and type of stress. For slopes
in the range 1:18 to 1:12 efficiencies in the order of 80 to 90 per cent
can be achieved. Investigations in the United Kingdom (29) have shown
that finger joints were at least as strong as timber having 50 per cent
clear material (50 grade) at any section.

It cannot be over emphasised that the preparation of glued joints requires careful supervision and the following comments are intended to
form a basis for good gluing practice. Firstly there are some basic
rules for adhesion:

(a)   the adhesive should 'wet' the timber and thus all surfaces to be
glued should be clean and free from dust, dirt, sawdust, oil and all other
contaminating substances. A surface is completely unsuitable for
gluing if the adhesive gathers in drops and does not spread and penetrate the adherend. It is desirable to glue surfaces as soon as possible
after planing, preferably within 48 hours. The sanding of previously

planed surfaces is not desirable as the timber surfaces can be coated
with dust which hinders good bonding. The effect of various surface
pre-treatments on the strength of glue lines is discussed Chugg (30)
in a very comprehensive text on gluing practice. With regard to plywood, adhesive manufacturers recommend sanding even if the surface
is free from contaminating substances. This is to remove the apparently glazed or microscopically smooth external surface of the plywood
which is thought to be due to the application of heat and pressure required in its manufacture. The term 'case hardening' is used to des-
cribe this non-gluing defect and is sometimes used incorrectly to des-
cribe machined timber which has been left for a considerable period
prior to gluing. The cut cells tend to close and the wettability of the
surface is reduced.

(b) The adhesive should be mixed and spread strictly in accordance
with the manufacturers data and at the correct temperature. Care must
be taken to ensure an even distribution of glue over the contact surface.
'Hit and miss' application is potentially dangerous as it may lead to
stress concentration and thus cracking of the joint.

(c) The moisture content of the timber must be carefully controlled
during gluing and should not exceed 15 per cent. If woods of differing
moisture content are glued together there will be differential shrinkage leading to stress concentration at the glue line. The moisture content of adjacent pieces of timber at a joint should not differ by more
than 6 per cent.

(c) A satisfactory glued joint cannot be achieved without pressure to
bring the surfaces into contact. This can be achieved by a suitable
nailing pattern (14) and for glued joints between plywood and solid
timber the maximum spacing of nails along the grain should not exceed:

100 mm for plywood not more than 10 mm thick
150 mm for plywood more than 10 mm thick

The length of round wire nails should not be less than four times the
thickness of the plywood in contact with the nail head. Side spacing
between lines of nails should not exceed 100 mm. To ensure the edge of
the plywood is tight against the solid timber an extra row of nails may
be required but the edge distance should not be less than five times
the nail diameter. Rippling of the plywood between the nail points is
an indication that the nail spacing is too large.

(d) Reference to Chapter 3 will indicate that there are a number of
synthetic resins commercially available for structural applications
whose properties can be modified over a large range. It is important
that the manufacturer should be aware of the conditions under which the
glued joint is expected to work satisfactorily. The suitability of some
formulations of urea formaldehyde, for example, are in doubt for permanent structures, even in dry conditions. Types of adhesive for use
with timber and their decline in strength over a period of time is discussed at length by Chugg (30). For exterior applications a Resorcinol
type resin adhesive is the natural choice as it can normally withstand

severe conditions of exposure, heat and humidity. There is little information available on the long term performance of glued joints but some tests carried out in Germany (31) have produced encouraging results for Resorcinol resin glues.

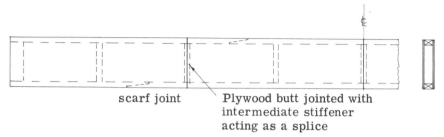

scarf joint   Plywood butt jointed with
                          intermediate stiffener
                          acting as a splice

Fig. 9. 40 The position of web splices in box beam construction should be staggered.

Returning to the design of joints for box and I sections, which are required in the chords and plywood webs, the structurally preferable arrangement of joints is not always compatible with that which is commercially the most economic. It is desirable to stagger the positions of web splices in box beam construction and nor should they coincide with the chord joints (Fig. 9. 40). Tests carried out by the authors (see Chapter 15) have indicated that for box beams of spans up to 9.0 m an adequate web joint can be achieved by using the internal web stiffener as a blocking piece behind the glue line. An alternative is to use a splice piece (Fig. 9. 41) and design guidance for this type of joint is

Fig. 9. 41 Butt joint with internal plywood splice plate.

given in reference (21). Finally it must once again be emphasised that glued joints are very susceptible to defects during manufacture and also wet and dry cycling in service may produce stress levels in excess of those assumed for normal loading conditions. The most refined structural calculations are useless if the designer gives inadequate consideration to the durability of the materials he is using. Wood is subject to two natural hazards, insect and fungal attack, which affect its durability. These and the problem of fire resistance are examined in the following two chapters.

10 Durability of timber

Two questions always occur when timber is being discussed as a permanent building material, one is its fire resistance, the other is its ability to withstand rot and infestation. The inference is that timber has two weaknesses—it will burn and it is not durable—and these are sufficient cause to preclude the consideration of timber as a permanent building material.

Regarding wood's durability, the most conclusive, if somewhat emotional, argument in its favour is to cite the many examples, in this country alone, of timber buildings, constructed in the Middle Ages and still structurally sound today. The point is that any building material is to some extent impermanent—brickwork crumbles, steel rusts. The secret of good design is to use all materials, being fully aware of their shortcomings, in such a way as not to expose their weaknesses. In order to be able to do this with timber, a full understanding is necessary of the conditions which can jeopardise its durability.

Wood, used in building, is subject to two natural hazards; one is infestation by wood-boring insects; the other is infection by wood-destroying fungi, otherwise referred to as rot.

INFESTATION BY WOOD-BORING INSECTS

The least serious of the two natural hazards in this country is that of insect attack. There is no doubt that if timber becomes infested and such infestation is allowed to develop unhindered, eventually wood can be reduced to powder. This process, however, takes a long time and it would usually need excessive neglect for the infestation to reach such a proportion as to cause collapse. Also, unlike rot, infestation, having

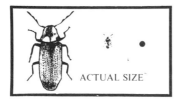

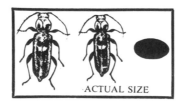

Fig. 10.1 Common furniture beetle, death watch beetle, and (below) the house longhorn beetle. Copyright: SGB Group of Companies.

occurred, can be brought under control by superficial treatment with chemical insecticides. All species of wood-boring insect go through parts of their life cycle on the surface of the timber—the laying of the egg, the hatching out of the larvae and the flight after the pupa stage— and are therefore vulnerable to contact insecticides.

Generally, insect infestation is encouraged by cool, damp conditions which create a congenial environment for the larvae; and it is the insects' larvae which cause the damage to the wood. Often the presence of fungal decay is conducive to infestation.

In this country the most usual insects to attack timber are the common furniture beetle, the house longhorn beetle, the death-watch beetle (Fig. 10.1) and the powder-post beetle. The first two beetles cause the most frequent trouble, although the first is far more widespread than the second. There are other insects that attack wood, but none of these is a danger to timber when used in building. They all represent a hazard to timber in its more natural state and their infestation dies out rapidly when timber is converted and seasoned.

Such harmless insects are:-

(1) The ambrosia beetle, which attacks logs in the forest.

(2) The marine borer or shipworm, which attacks logs as they are floated down from the lumber camp to the sawmills and which die when the logs are removed from the water.

(3) The woodwasp, which attacks trees or logs, but which dies out a year after the timber is converted.

(4) The forest longhorn and jewel beetle, which attacks trees and logs and which dies out after a few years. This beetle must not be con- fused with the more dangerous house longhorn beetle.

(5) The waney-edge borer, which lives below the bark of logs and newly converted timber and which dies out after a year or two.

(6) The powder-post beetle (*Bostrychidae*), which attacks sapwood of imported hardwoods and dies out within one year. This beetle must not be confused with the powder-post beetle (*Lyctidae*).

(7) The wood-boring weevil (*Euophryum confine*), which is similar in appearance to the furniture beetle, but is associated with wood that is damp and rotted. Measures to deal with the rot will eliminate the weevil infestation.

So much for insects which, while attacking timber, are not harmful to timber in building. Now for the ones that can infest wood after it has been used in structures. All four insects have many features in common. In each case it is the larvae of the insect that attacks the wood, continuously boring its way through it during this stage in its development, feeding on the starch it contains and leaving behind a network of tunnels. Hence the common reference to wood-worm attack. However, these grubs are not worms; they are merely larvae, which eventually pupate and then turn into a winged insect.

COMMON FURNITURE BEETLE (*Anobium punctatum*) (Fig. 10.2)

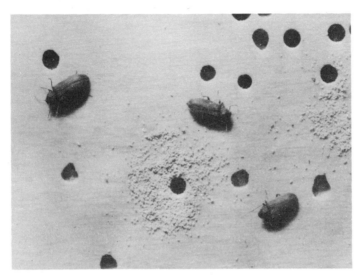

Fig. 10.2 Furniture beetle. Copyright: SGB Group of Companies.

Over 75 per cent of all damage by wood-boring insects in this country can be attributed to the common furniture beetle. It is indigenous to the great mass of temperate Europe and attacks furniture, joinery, structural timbers and plywood made from less durable timbers, such as birch and alder and with glues of a natural origin.

Sapwood (the outer layers of wood in the trunk of a tree, recognisable by its lighter colour) is liable to attack by the furniture beetle whether of soft or hardwoods. Heartwood, however, is rarely attacked.

The life cycle of the furniture beetle is similar to other varieties of wood-boring insects and to understand preventive and remedial treatments, it is necessary to understand this cycle.

The female beetle lays its eggs (up to 80 in number) in crevices on the unpainted surface of timber. In about three weeks these hatch out and the grubs bore into the wood, where they continue to live for two or more years, boring a maze of tunnels, which eventually can cause the disintegration of the timber. In time the larvae pupate in a small cavity near the surface of the wood. In two or three weeks (usually in the spring or early summer), the adult beetle emerges and makes its exit from the wood through a flight hole about 1.6 mm in diameter. It is the evidence of these flight holes and the little tell-tale pile of wood dust associated with them that draws attention to a 'live' infestation.

HOUSE LONGHORN BEETLE *(Hylotrupes bajulus)* (Fig. 10. 3)

Fig. 10. 3 House longhorn beetle. Copyright: SGB Group of Companies.

This beetle has a very restricted area of operations at the present time and is found mainly in the northern parts of Surrey. The confined nature of its habitat has led to the actual local authorities in whose area they exist being listed in the Building Regulations 1965. All structural timbers of buildings in these areas are required to be protected by the regulations and the 'deemed to satisfy' clauses associated with this requirement suggests immersion of all timbers for 10 minutes in an organic solvent containing at least 0. 5 per cent of a contact insecticide of the chlorinated hydrocarbon type.

The Building Regulations make no reference to other forms of insect attack, which suggests the extent of the damage the house longhorn beetle can cause. The size of its oval flight hole (10 mm × 5 mm) is an indication of the more extensive damage of which it is capable. Like the furniture beetle, the house longhorn beetle attacks sapwood of softwoods and infestation can eventually persist until the sapwood disintegrates. Larvae can live from four to seven years in the wood before pupating and adult beetles can lay as many as 200 eggs.

DEATH-WATCH BEETLE *(Xestobium rufovillosum)* (Fig. 10. 4)

This infestation usually starts in timbers which have already become decayed and is generally restricted to hardwoods, both sapwood and heartwood. The beetle's flight hole is circular and of about 3 mm in

Fig.10.4 Flight holes of the Death Watch beetle.
Copyright: S.G.B. Group of Companies.

diameter and the larvae spend from three to ten years in the timber.
The well-known tapping noise associated with the death-watch beetle is
caused by the head of the adult beetle and is a mating call during the
flight season. The female lays up to 200 eggs.

POWDER-POST BEETLE *(Lyctidae)*

This beetle must not be confused with the powder-post beetle imported
in the sapwood of certain hardwoods. Lyctidae, sometimes called the
parquet Beetle, infests the sapwood of home-grown hardwoods and will
cause, if allowed to proceed untreated, the disintegration of this wood
leaving a thin layer of sound timber on its surface. It is slightly larger
than the furniture beetle and its flight hole is circular and 1.6 mm in
diameter, smaller than its imported relation which has a 5 mm flight
hole. Lyctidae usually attacks newly converted timber in sawmills and
factories and has a short life cycle of only about ten months.

PREVENTATIVE AND REMEDIAL TREATMENT

It is possible to give timber protection against future infestation by all
wood-boring insects. The principle of such treatment is to impregnate
the components, either under pressure or by immersion, with an insecticide before the timber is assembled or built into the structure. The
insecticide produces a toxic layer of timber on the outside of the components, which either kills by contact the eggs, or the beetles laying the

eggs or emerging from the timber, or the larvae tunnelling near the surface by ingestion.

Two processes available are also used as preservative treatments against fungal attack, and will be dealt with in more detail later.

These are either vacuum/pressure impregnation with copper, chromium and arsenic salts, or diffusion with boron salts. Alternatively, components can be treated by dipping in a penetrating organic solvent solution of chlorinated hydrocarbon type of contact insecticide (at least 0.5 per cent solution) such as dieldrin or lindane (gamma BHC).

The degree of penetration of this insecticide into the wood is an important aspect of the treatment and care must be taken not to lose the toxic layer by later machining.

Remedial treatment, where an infestation has been discovered, can be undertaken by a brush or spray application of contact insecticide, preferably timed before the adult beetles leave the wood. The degree of penetration of the insecticide is obviously slight, but because of those parts of the beetles' life cycle which take place on the surface of the wood, such treatment, if undertaken thoroughly and including all surrounding timbers to those affected, can be quite effective. Lindane can also be deposited by smoke generation. This gives no penetration and therefore may well require repeated treatments at yearly intervals before the infestation is eradicated.

But how serious is insect attack? Infestation can be ignored completely on external joinery, as indeed it can on most completely painted timber assemblies. Timbers that are most likely to be attacked are those in roofs and suspended floors. In addition, statistics produced by the British Wood Preservers' Association in 1967 suggest that the incidence of infestation and decay (other than dry rot) is markedly greater in country districts. In fact, only 27.6 per cent of cases of defect occurred in towns and cities. The higher incidence in rural districts was attributed to infection from adjacent woodlands and hedgerows.

It would seem that periodic inspection of visible timbers and normal care in detailed design to avoid the condition conducive to rot, which can encourage infestation, can normally control insect attack. There does not seem to be a good case for general treatment of timber, unless the building in question is in house longhorn beetle country, where treatment is mandatory.

INFECTION BY WOOD-DESTROYING FUNGUS

The question of fungal attack is a much more serious one. There is, however, one fact which is incontrovertible and which makes the whole question of fungus control possible. Wood destroying fungi only attack timber when its moisture content is above 20 per cent and timber in building should normally have a moisture content no greater than 14 per cent. Therefore, certain assurance against infection is thoughtful design and the selection, as far as possible, of those types of wood which are more resistant to rot for use in vulnerable situations. Timber is particularly at risk where it is in contact with the ground, at or below damp proof course level, built into brickwork or embedded in concrete, in unventilated situations, or in those areas where the normal moisture, either from external sources or from condensation, is likely to encourage the timber to take up a moisture content of 20 per cent or over. There is nothing inherently dangerous in timber getting wet, so long as it is allowed to dry out within a reasonable period.

TABLE 10.1 Table of durability

Type of Timber	Approximate life in contact with ground (years)
Imported Softwoods:	
Douglas Fir	10-15
Pitch Pine	10-15
Western Hemlock	5-10
Parana Pine	5-10
Redwood	5-10
Whitewood	5-10
Canadian Spruce	5-10
Western Red Cedar	15-25
Home-grown Softwoods:	
Douglas Fir	10-15
European Larch	10-15
Scots Pine	5-10
European Spruce	5-10
Sitka Spruce	5-10
Imported Hardwoods:	
Africa Mahogany	10-15
Afromosia	more than 25
Greenheart	more than 25
Troko	more than 25
Meranti	10-15
Sapele	10-15
Teak	more than 25
Home-grown Hardwoods:	
European Ash	less than 5
European Beech	less than 5
European Oak	15-25

All timber is to some extent liable to fungal attack if the conditions are conducive, but some types are more resistant than others. Sapwood of many species is particularly vulnerable; heartwood is normally resistant to rot. The most commonly used timber for building in this country is Baltic Redwood (Pinus Sylvestris). Redwood's sapwood is liable to

rot; its heartwood is much less vulnerable. It would, however, be econo-
mically impossible to use exclusively heartwood. In any batch of redwood, there will be a certain proportion of sapwood, and for this reason
good detailing, joinery practice, careful site storage of timber, good
painting specifications, reasonable maintenance of timber in buildings
are all essential. If these practices are all observed without exception,
there would be little danger of rotting timber. However, as there is
always the possibility of oversight, design miscalculation or lack of
maintenance, it is better to pre-treat all timbers with a preservative,
or at least, those timber components in particularly critical situations.

Different timbers have different degrees of resistance to rot. Western
red cedar and douglas fir, for instance, have a good resistance and
preservation is not required. Whitewood (Spruce) and hemlock, on the
other hand, have low durability and need preservative treatment where
used in dangerous areas. These two species have, however, low permeability and therefore pressure impregnation treatment should be
used rather than an immersion one. Ash and Beech will decay in one to
five years, if used in contact with the ground or in damp conditions;
Afromosia, Iroko and Teak, though, in similar conditions will remain
relatively unharmed for 30 years or more. The durability of different
timbers is illustrated in Table 10.1.

It was because of the vulnerability of certain timbers to rot that the
National House-Builders Registration Council have required, since 1969,
all external joinery in buildings covered by their guarantee to be
treated.

It should be remembered that there are many timber buildings still as
good today as the day they were built in the Middle Ages. In those days
there were no preservative treatments and it was merely the careful
design of the structures that protected the timber from the causes of
harm. This coupled with the selection of appropriate timber for particular uses, has preserved these buildings to the present day.

TYPES OF FUNGAL ATTACK

There are many fungi which will live on timber, but certain fungi are
not, from the technical point of view, harmful to the timber. They do not
live on the wood itself, but on substances that are the remains of the sap
circulation when the wood was growing. These fungi can discolour wood,
but do not cause it to disintegrate. Such a fungus is the blue fungus often
seen on pine sapwood.

The really harmful fungi, those that cause timber to break down, are:

    dry rot
    cellar fungus
    mine fungus

Dry rot (*Merulius lacrymans*) (Fig. 10. 5)

Fig. 10. 5 Dry rot fungus. Copyright: S.G.B. Group of Companies.

Dry rot attacks timber in damp and unventilated situations. The spores of the fungus are carried in the air and when they alight on a suitable timber of the right moistness they start to grow. Cotton-wool like growths form and, when once established, a netwood of fine threads (or hyphae) are sent out, searching for more wood to attack. The fungus extracts nourishment from the timber, breaking down the cellulose to carbon dioxide and water and quickly reducing the wood to a brittle condition. Cuboidal cracking of the timber is typical of this type of attack.

Dry rot particularly attacks softwoods, but can spread to hardwoods. Once the fungus is established the moisture of respiration of the fungus itself helps to perpetuate the conditions in which it can flourish.

Cellar fungus (*Coniophora cerebella*) (Fig. 10. 6)

Cellar fungus, or wet rot, is most common in external timbers, such as gate posts, garden sheds etc., but it is sometimes found under floor coverings on damp floors. It may attack soft or hardwoods and requires more moisture than dry rot. Cellar fungus often leaves a sound surface veneer of wood while completely destroying the timber beneath. This makes its detection sometimes difficult. Unlike dry rot this fungus does not form a fleshy fruit body but a thin, olive green, flat sporophore.

Fig.10.6 Cellar fungus.   Copyright: S.G.B. Group of Companies.

Mine fungus *(Poria vaillantii)*

Mine fungus is usually found in damp mines, but it can be present in buildings, where its appearance is much like dry rot. A close relative of this *Poria vaporaria* which is another wet rot having a characteristic fan shaped spread to its white hyphae.

In the case of all fungal attacks, when discovered remedial action must be urgent and thorough. Infected timber should be cut out and destroyed. All surrounding structure, whether timber or not, should be treated with a fungicide and, most important of all, the causes of the conducive conditions that encouraged the rot must be rectified.

PRESERVATIVE TREATMENTS

The criteria for good preservative treatment are that the greatest depth of penetration of the preservative must be achieved and it must dry out quickly leaving the timber undistorted and in a condition to be effectively glued and painted.

It is the requirement for deep penetration that makes brush applications of preservative of little use.

Great care must be taken in selecting the particular treatment, having in mind the eventual use of the treated components and the types of timber involved. Some timbers are easily permeated, such as Beech, Ramin, Baltic Redwood, Scots Pine and Birch; some are resistant to treatment, such as Spruce and Hemlock.

Permeable timbers can be preserved satisfactorily by immersion processes, but impermeable timbers will require pressure impregnation. Also it must be borne in mind that the preserved layers of timber (the outer layers) must not be lost by later machining. Therefore, with the exception of diffusion treatment which is undertaken at the sawmill, treatments should preferably take place after machining. This is essential in the case of immersion treatment, where the penetration is less deep. Treatment applied to finished components must not, therefore, alter the dimension of the component. Occasionally completed assemblies are treated. In this case the glue and glueing techniques must be such that the joints will not later open in use, thereby exposing untreated wood.

Methods of preservation treatment are:
    vacuum/pressure impregnation
    double vacuum impregnation
    immersion treatment
    diffusion

Vacuum/pressure impregnation

This is known as the full cell process and achieves the complete impregnation of the sapwood. Water-borne preservatives based on copper, chromium and arsenic salts are used and their composition and application are covered by BS 4072 : 1966.

The process requires complex equipment, but produces a treatment good for a life of 50 to 60 years or more, and the chemicals will not leach out in use. When completely dry, paint can be applied to the treated timber without problem.

This process has one major objection. The treatment causes the wood to swell and then shrink again on drying. This often results in the grain rising and therefore the process is not suitable for finished components. Trade names involved in this type of treatment are:

    'Celcure A'   (Rentokil Laboratories Ltd)
    'Tannalith C'   (Hickson's Timber Impregnation Co (GB) Ltd)
    'Treatim CCA'   (Treatim Ltd)

Pressure impregnation (the empty cell process) can be used with creosote (BSS 144 : 1954 and BSS 913 : 1971) but this is only suitable for large industrial uses such as fencing, marine piling poles, wooden bridges etc., and is of little application in building

Double vacuum impregnation

This process provides complete penetration of the sapwood, using a low viscosity organic solvent. The preservative is expensive; hence the double vacuum, the first to drive the preservative into the cells of the timber, the second to draw out excess preservative still filling the cells. In this manner controlled absorption can be obtained. The plant is more simple than that required for vacuum/pressure impregnation and the process is applicable for machined components or assembled joinery. After treatment the timber needs to dry for one or two days before being primed. Penetration of preservative is about twice that obtained in a three minute immersion process. The 'Vac-Vac' process of Hickson's Timber Impregnation Co (GB) Ltd and Protim Prevac of Protim Ltd are examples of the double vacuum process.

Immersion treatment

In this process the equipment can be extremely simple, being merely a tank of organic solvent preservative like that used in the double vacuum process, with or without the addition of a water repellent. The treatment is suitable for machined components or assembled joinery, but because of the lesser penetration than previous processes, no machining must be undertaken after treatment. A three minute dip is recommended (American experience suggests that this gives a twenty five year efficacy) but this can be reduced to one minute, with a consequent reduction in safety.

The average take-up of preservative solution on lateral surfaces of redwood sapwood in a three minute dip is about 185 g per m^2. Penetration can obviously vary considerably depending on the permeability of the individual pieces of wood. Average lateral penetration is 6.5 mm, but it can vary from 1 mm to 15 mm or more. Through end-grain, penetration normally extends from 40 mm to 80 mm. The minimum degrees of penetration really satisfactory are 3 mm on lateral surfaces and 40 mm on end grain surfaces.

Other methods of application can be used such as a coarse spray tunnel or machine deluging. Both give the effect of a one minute immersion.

Timber treated by immersion will require a drying period of between ten minutes and twenty four hours before priming.

Diffusion treatment

This is a sawmill application as only freshly sawn timber in an unseasoned state can be treated. To the timber is applied a water-soluble boron compound after which it is closely stacked to allow the preservative salt to diffuse before the timber is dried.

After drying, the seasoned timber is supplied to the joinery manufacturers just like untreated timber, but is in fact, thoroughly penetrated with preservative However, because of the water-soluble nature of the salts, there is a danger of them leaching out if the timber is exposed to rain for long periods. As a result, all timber must be well primed before delivery to site.

Painting can take place on treated timber, when dried, without trouble.

The trade name associated with diffusion treatment is 'Timborised' timber. The process is operated in accordance with BWPA Provisional Standard 105 by Borax Consolidated Ltd and timber so treated comes into this country from Canada, Finland and Russia.

Organic solvent preservatives, as used in the double vacuum and immersion processes, have the advantages of not distorting finished components and drying out quickly. They are, however, expensive. Water repellents are sometimes included. They result in the timber in use being less liable to swell and shrink and therefore forming more stable joinery assemblies with less consequent cracking of painted surfaces. Water repellents, though, tend to produce their own problems. A longer drying time is necessary before the application of paint. Repellents can sometimes reduce the adherence of the paint film and interefere with the drying and appearance of the finishing coat. Wax-based repellents are more satisfactory. Some sapwoods (a very small proportion, probably about 5 per cent) have patches of abnormal permeability and these can cause trouble to the paint application, particularly when a water repellent is used.

Proprietary organic solvent preservatives usually contain one of the following ingredients dissolved in a quick drying solvent:

(1) Pentachlorophenol in a 5 per cent solution, or a 4. 5 per cent pentachlorophenol solution plus copper or zinc naphthenate equivalent to a metal content of not less than 0. 4 per cent
(2) Lauryl pentachlorophenol in a 5 per cent solution
(3) Tributyltin oxide in a 0. 5 per cent solution
(4) Copper naphthenate solution containing 2. 75 per cent copper.

It should be noted that the vacuum/pressure process with copper, chromium and arsenic salts and the diffusion process with boron salts provide an insect protection as well as a fungal protection. The organic solvent treatments can also be provided with an insecticide additive to provide all round protection.

In conclusion, it should be emphasized that the above treatments should be considered merely as an additional safeguard, and not a total protection against fungal attack. Good design and detailing are the real safeguards. A full understanding of the causes of fungal attack and avoidance in the design of those conditions conducive to infection is the surest protection. If this is done, together with careful site management during building and intelligent maintenance of the completed building, timber can be considered in every way a permanent building material.

11 Timber and fire

Timber and fire are two words which, placed in conjunction, become so charged with emotion, that logical thought processes are sometimes forsaken.

The Great Fire of London in 1666 still lurks in all our school days' memories. The well meaning Lord Mayors of London in the twelfth century, who enacted regulations to restrict the erection of too many timber buildings in the city, foresaw the problem and endeavoured to do something about it. Had it not been for the failure to enforce these regulations, the Great Fire might have been avoided and the subject of timber and fire would have been a less emotive one.

The point really is that fire hazards invariably arise from the contents of a building and not from the building structure itself. For example, in 1967 there were 95, 447 fires to which fire brigades were called. In the subsequent reports, floors, skirtings, stairs and roof members are mentioned in 2, 730 cases as the materials first ignited (or 2.86 per cent of the total); while a further 407 times timber is mentioned associated with a chimney and 814 times below a hearth (both cases of faulty building construction). Adding these figures to the earlier one, we find that only 4 per cent of the fires were associated initially with a timber element of the building.

Once the fire has started, it is propagated by the multitude of combustible and easily ignitable objects we surround ourselves with in our daily lives. Only then, when the room in question is a roaring inferno, does the building structure usually contribute to the propagation process. How soon, and with what ferocity it does so, is all a question of design.

In the case of the Great fire of London, once the fire had started at the pastry cook's, it was propagated rapidly through the thatch roofs of the city, aided and abetted by the combustibility of the vast quantities of timber in the too-closely packed buildings.

Today, insurance companies, although rapidly modifying their traditional opinions in the light of present technology, view timber with some concern. Building Societies have the same prejudices. Their position is understandable in the light of cheering memories of crackling log fires. The point is that timber *is* combustible; but is not easily ignited, will not ignite itself and equally has quite special qualities which render its performance in a fire predictable and not unhelpful to the fire brigade.

COMBUSTIBILITY

Combustion is defined by the Oxford English Dictionary as 'the development of light and heat accompanying chemical combination'. A non-combustible material (according to BS 476 which controls the fire testing of building materials) is 'one which neither burns nor gives off inflamable vapours in sufficient quantities to ignite at a pilot flame when tested in the manner defined . . . '.

Timber is combustible. Such inorganic materials as brick, concrete, asbestos, metals and glass are classed as non-combustible (in accordance with the appropriate test in BS 476 Part 4 1970), but no material is entirely fire-proof or unaffected by the kind of temperatures engendered in a burning building (700°C to 900°C). In fact, some incombustible materials are subject to quite dramatic failure due to high temperatures; concrete can disintegrate under the effects of heat; some natural stones even shatter; and steel loses half its strength at about 500°C.

Timber, on the other hand, has certain very special characteristics. Owing to its very low thermal conductivity, due to its cellular structure, timber resists the spread of fire to some extent. What is more, its likelihood of failure is entirely related to the distance the fire has eaten through the timber, and the unaffected timber inside the section will remain strong and not collapse due to the effect of heat, as would a metal member. In short, timber will not be subject to dramatic failure in the event of fire (Fig. 11. 1).

The graph (Fig. 11. 2) produced by TRADA, shows the comparative fire resistance of two sizes of timber members, a mild steel member and a

Fig. 11. 1 The characteristics of timber & steel in a fire.
Copyright: CIBA-GEIGY (UK) Ltd.

typical aluminium alloy member. The performance of these materials is set against a time-temperature curve reproduced from BS 476. From this it is seen that the level of assumed failure (15 per cent of initial strength) was reached by the various materials, in the following times:

Curve C	Aluminium Alloy	4 minutes
Curve B	Mild Steel	9. 5 minutes
Curve D2	Timber 25 mm × 50 mm beam	13. 5 minutes
Curve D1	Timber 25 mm × 50 mm tie	14. 5 minutes
Curve E2	Timber 50 mm × 100 mm beam	24. 5 minutes
Curve E1	Timber 50 mm × 100 mm tie	26. 5 minutes

The differences between the steel example and the first two timber examples are hardly significant. It does serve to illustrate though that 25 mm thick timber has at least the equivalent fire resistance of unprotected steel. A thicker timber member has very much superior characteristics.

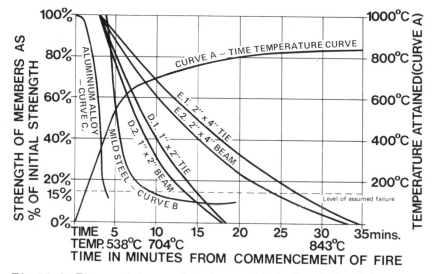

Fig. 11.2 *Fire resistance of timber, mild steel, and aluminium alloy members (graph produced by TRADA)*

The Timber Research and Development Association quote a test of a glulam and a steel beam, both of equivalent strengths, supporting identical loads over a chamber in which there was sited a controlled heat source, producing a standard time-temperature curve. Their reactions showed remarkable variations. The steel beam started to sag within minutes. After 29 minutes it had a 900 mm deflection and collapsed one minute later. Meanwhile the glulam beam had deflected only 56 mm while 75 per cent of its cross sectional area was unaffected by fire. To understand timber's curious characteristics, the way wood burns needs to be examined.

How wood burns

When wood is heated, its temperature rises steadily to 100°C, when its water content is vapourised and driven off as steam, or pushed from the surface layers of wood into the centre of the member. The vapourisation causes a temporary check in the temperature rise. When the rise recommences, a chemical change starts in the wood and disintegration begins. Below 200°C there is a darkening of the wood and between 250°C and 300°C the formation of charcoal on the exposed surfaces is evident, together with the giving off of inflammable gases, including carbon dioxide and methane. It is the ignition of these gases which causes wood to flame. After flaming is over 'smouldering' or glowing commences. This is the combustion of the charcoal and the formation of ash. This process can only start when the temperature reaches about 500°C. Until that temperature, the charcoal remains more or less inert, protecting the timber beneath. Flaming is largely responsible for fire spreading and smouldering for sustaining the fire.

When wood starts to burn and moisture is driven into the centre of the timber, its moisture content is raised. As a result, the strength decreases sharply for a short time, but then further effects are negligible. This means that timber largely retains its strength, having taken into account the reduction of the cross sectional area that results from the charring process.

It is wood's predictability both in its strength-retaining characteristics during a fire and its rate of burning that make it a unique material. In spite of being combustible, it will not self-ignite, even after prolonged heating to temperatures below 100°C. In addition, timber is not easily ignitable when exposed to a small heat source as described in BS 476: Part 5:1968.

Timber has a predictable charring rate—for average timbers in the order of 0.6 mm per minute. Certain hardwoods, such as teak or greenheart, have a slower rate; some less dense species char more

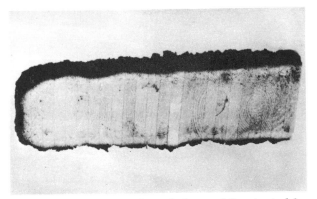

Fig. 11.3 Cross section of charred laminated beam.
Copyright: CIBA-GEIGY (UK) Ltd.

rapidly. In the case of free-standing columns, the rate would be increased by a factor of 1.3 as the timber is exposed on all four sides. This consistency of performance also applies to laminated members made with thermosetting synthetic resin adhesives (Fig. 11. 3). It is, in fact, possible in the case of substantial timber components, to allow in the size of the member for a period of fire protection required; that is after the 'sacrificial' timber is charred, the member will still not collapse. For this purpose, collapse stresses are taken as 2. 25 × safe working stress for beams and 2 × safe working stress for columns.

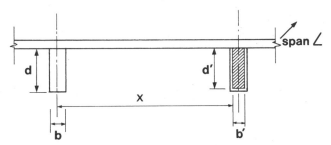

Fig. 11. 4 Domestic floor with solid timber joists.

The following example illustrates how the concept of sacrificial timber can be used in calculations to assess the fire resistance of timber beams. A similar procedure may also be applied to columns. The derivation of equations used in this example is given in chapter 9. Consider a domestic floor with solid timber joists of span L metres and spacing x metres (Fig. 11. 4). The proportions of joists of this type will normally be governed by deflection requirements and if a limit of 0. 003L mm is imposed on the maximum deflection under uniformly distributed load, the total load W^{KN} that can be carried by the beam is given by

$$W = 0.0192 \frac{Ebd^3}{L^2} \times 10^{-9}$$ 11(i)

where E is the modulus of elasticity of timber N/mm^2 and the remaining notation as in Fig. 11. 4. For a given width b mm and depth d mm of section the maximum working load can be determined from equation 11(i). The section must now be checked to ensure that the permissible bending and shear stresses are not exceeded. Using the value of W obtained from equation 11(i) the maximum shear and bending stresses may be obtained from equations 11(ii) and (iii) respectively which should not exceed the permissible values.

$$\text{Shear stress } q^N/mm^2 = \frac{3W}{4bd} \times 10^3$$ 11(ii)

$$\text{Bending stress } f^N/mm^2 = \frac{3WL}{4bd^2} \times 10^6$$ 11(iii)

Putting some typical figures into equation 11(i) ($L = 4$ m, $b = 50$ mm, $d = 200$ mm, $E = 8400^N/mm^2$) the permissible load is given by

$$W = \frac{0.0192 \times 8400 \times 50 \times 200^3}{4^2 \times 10^9}$$

$$= 4.03 \text{ KN}$$

If the joist spacing $\times = 0.4$ m then the maximum uniformly distributed load w KN/m^2 is given by

$$w = \frac{W}{\times.L} = \frac{4.03}{0.4 \times 4} = 2.5^{KN}/m^2$$

The shear and bending stresses corresponding to the load $W = 4.03$ KN are given by

$$q = \frac{3 \times 4.03 \times 10^3}{4 \times 50 \times 200} \quad \text{from 11(ii)}$$

$$= 0.302 ~^N/mm^2$$

$$f = \frac{3 \times 4.03 \times 4 \times 10^6}{4 \times 50 \times 200^2} \quad \text{from 11(iii)}$$

$$= 6.06 ~^N/mm^2$$

Typical design stresses for the grade of timber used for domestic construction are bending $6.7^N/mm^2$ and shear $0.77^N/mm^2$. These values would be appropriate to a 50 grade Redwood (see chapter 7) and thus the calculated values are acceptable.

Suppose the 50 mm $\times$ 200 mm joists are exposed to fire for a period of say 15 minutes on the soffit and two sides and the charring rate is 0.64 mm/minute. The reduced width

$$b^1 = b - 2 \times 0.64 \times 15$$

$$= 50 - 19.2$$

$$= 30.8 \text{ mm}$$

and the reduced depth

$$d^1 = 200 - 0.64 \times 15$$

$$= 200 - 9.6$$

$$= 190.4 \text{ mm}$$

In the event of a fire deflection is not critical and the failing load at a given time interval may be estimated by rearranging equations 11(ii) and (iii) using the appropriate characteristic stresses. In Chapter 5 it was indicated that the characteristic stress is equal to the design stress multiplied by a safety factor of 2.25.

Thus equations 11(ii) and (iii) may be written as

$$\text{W ultimate} = \frac{4\,(2.25q)\,b^1 d^1}{3.10^3} \text{ for shear} \qquad\qquad 11(iv)$$

$$\text{W ultimate} = \frac{4\,(2.25f)\,b^1 d^{1^2}}{3.L.10^6} \text{ for bending} \qquad\qquad 11(v)$$

where q and f are the design stresses for shear and bending respectively. For $q = 0.77^N/mm^2$ equation 11(iv) gives

$$\text{W ultimate} = \frac{4 \times 2.25 \times 0.77 \times 30.8 \times 190.4}{3000}$$

$$= 13.55 \text{ KN}$$

and for f $= 6.7^N/mm^2$ equation 11(v) gives

$$\text{W ultimate} = \frac{4 \times 2.25 \times 6.7 \times 30.8 \times 190.4^2}{3 \times 10^6 \times 4}$$

$$= 5.61 \text{ KN}$$

These results show that after a period of 15 minutes the effective timber section 30.8 mm × 190.4 mm (Fig. 11.4) is adequate to carry the normal design load of 4.03 KN. After a further period of 15 minutes the effective timber section will be

$$b^1 = 50 - 2 \times 0.64 \times 30$$

$$= 50 - 38.4$$

$$= 11.6 \text{ mm}$$

$$d^1 = 200 - 0.64 \times 30$$

$$= 200 - 19.2$$

$$= 180.8 \text{ mm}$$

The effective width of the section is now less than 25 per cent of the original 50 mm width and W ultimate for bending is 1.9 KN. Thus the section does not have a fire rating of 30 minutes without protection. It can be seen from the above calculations that where both sides are subjected to charring thin sections will have a low fire rating. If in the above example the width b was increased to 75 mm then after a 30 minute fire with $b^1 = 75 - 38.4 = 36.6$ mm and $d^1 = 180.8$ a 30 minute fire rating could be achieved as

W ultimate shear $= 15.28$ KN
W ultimate bending $= 6.01$ KN

For large span laminated timber roof beams a fire rating in excess of 30 minutes may readily be achieved. Consider 115×500 mm laminated beams at 2.4 m centres designed to carry a total load of $1.5^{KN}/m^2$ over a span of 10.3 m.

The total design load W $= 1.5 \times 2.4 \times 10.3$

37.08 KN

For a 30 minute exposure to fire on the soffit

$b^1 = 115 - 38.4 = 76.6$ mm

$d^1 = 500 - 19.2 = 480.8$ mm

for a design shear stress of $1.36^N/mm^2$

$$\text{W ultimate} = \frac{4 \times 2.25 \times 1.36 \times 76.6 \times 480.8}{3000}$$

$$= 150.26 \text{ KN}$$

and for a design bending stress of $11.8^N/mm^2$

$$\text{W ultimate} = \frac{4 \times 2.25 \times 11.8 \times 76.6 \times 480.8^2}{3 \times 10^6 \times 10.3}$$

$$= 60.8 \text{ KN}$$

Thus a 30 minute fire rating is easily achieved under full design load of 37.08 KN.

In all questions of the burning of timber it must be remembered that fire attacks at the weakest places, namely the arises and joints. Sheets of plywood are more fire resistant than tongued and grooved boarding where the thickness of the tongue is the critical item. If the joints of plywood decking are fixed over joists, a 12 mm thick plywood sheet is equivalent to 19 mm tongued and grooved boarding, 15 mm plywood to 25 mm boarding. In fact in North American solid plank construction, where 100 mm thick tongued and grooved softwood planks are commercially available, high degrees of fire resistance are obtainable, depending on the thickness of the planks and the efficiency of the jointing. Fire resistant periods of two hours can be achieved.

One thing, however, must be remembered, the use of metal fixing devices, such as bolts, can speed the charring process due to the greater transmission of heat through the metal and the consequent attack of timber deep in the cross sectional area of the member.

TIMBER, THE BUILDING REGULATIONS AND FIRE

The effect of the 1965 Building Regulations was to make the use of timber in certain circumstances more acceptable than under the old bye laws. The section of the Building Regulations dealing with fire is Part E, 'Structural Fire Precautions' and its provisions with regard to the use of timber fall broadly under the following headings: fire resistance of elements of structure; spread of flame within the building; external wall treatments; external fire resistance to roofs.

Fire resistance of elements of structure

Unlike the old bye laws, there is no general requirement for the use of non-combustible materials in the Building Regulations, with certain exceptions.

(1)   Compartment walls and floors in buildings requiring one hour or more fire resistance.
(2)   External walls in buildings close to a site boundary or over 15.240 m high (unless they are far enough from the site boundary to be 100 per cent unprotected)
(3)   Walls around stair wells (Scottish Building Regulations only)

It will be seen, therefore, there is no general barring of the use of timber because it is a combustible material, so long as the relevant fire protection is given to the timber either by linings or sacrificial timber to give the structure validity for the required period stated in the Regulations. This period varies with the purpose and size of the building in question.

Basically buildings no higher than 7.620 m may have structures of timber, unless they are 'institutional' with disabled, sick or very young occupants. There are, however, known waivers for this condition. Non-residential buildings may have timber frames up to 15.240 m height with certain reservations. Flats and maisonettes up to four storeys have been built with timber compartment floors, but with non-combustible compartment walls. There was even a recent waiver to allow the TRADA to build a four storey block of maisonettes with timber compartment walls clad in a non-combustible material.

The periods of fire resistance to the timber structures are usually obtained by plasterboard or asbestos insulation board linings. Twelve millimetres of plasterboard gives a fire protection for half an hour; twelve millimetres of asbestos insulation board one hour. Even wood chipboard provides a fire resistant lining (19 mm thickness gives half an hour protection).

No fire protection is required for most single storey buildings except for compartment or separating walls. Non-loadbearing wall panels are also exempt, except when close to the site boundary. For the purpose of this section of the Regulations the roof and non-loadbearing partitions are not considered elements of structure and therefore require no fire protection, (except in the case of the roof which needs resistance to fire from outside).

Spread of flame within the building

All internal linings have to comply with a classification system as defined in BS 476:1970. The classifications are:-

Class 0   non combustible linings having a maximum of 0.794 mm covering of combustible material. (Such linings would be plaster, plasterboard, asbestos, brick, concrete etc.)

Class 1   linings giving a surface spread of flame of 165 mm after $1\frac{1}{2}$ minutes exposure to a standard flame

Class 2   linings giving a surface spread of flame of 215 mm after $1\frac{1}{2}$ minutes exposure to a standard flame

Class 3   linings giving a surface spread of flame of 265 mm after $1\frac{1}{2}$ minutes exposure and a final spread of 710 mm. (Most untreated timber lining products fall into this category).

Class 4   linings giving a surface spread of flame greater than Class 3. These are not acceptable for building purposes.

A table of location is laid down in the Building Regulations where the various classified materials can be used for walls and ceilings. Broadly, in all buildings other than small domestic ones, circulation areas must be lined in Class 0 materials. Otherwise Class 1 materials must be used, except in those rooms defined as 'small rooms' by the Regulations—the size of the 'small room' varying with the purpose group of the building. The exception to this is Institutional buildings where Class 1 material is required throughout, except for walls to 'non-small rooms' and, of course, circulation areas, which must be Class 0.

All timber is inherently Class 3, except Western Red Cedar and Balsa which fall into Class 4. This is because these timbers have a lighter cellular structure. Timber, plywood and chipboard can be given a Class 1 rating by impregnation with fire retardants or by surface coating (see later). Chipboard and plywood can be treated in manufacture to give Class 1 characteristics.

External wall treatments

The Building Regulations place restrictions on the use of combustible linings on the exterior of certain walls close to site boundaries. Except on buildings over 15.240 m high or within 914 mm of the boundary, it is allowed to use timber claddings not less than 9.52 mm thick. However, in relation to regulations dealing with the permissible 'unprotected' area of a wall (wall, including windows, without a fire resistance) and its proximity to site boundaries; timber cladding fixed to walls having internal fire resistance results in 50 per cent of the area of cladding being considered as unprotected wall area.

External fire resistance to roofs

The Building Regulations refer to BS 476:Part 3:(1958) which set standards required regarding the penetration of the roof by fire from an adjoining property and methods of testing this characteristic.

Suffice it to say that timber support frames with plywood decking with normal roof coverings is deemed satisfactory for these requirements.

The above is not supposed to be a comprehensive examination of the Building Regulations regarding timber and fire, but is merely a short appraisal of timber as its use is affected by the Regulations. It is obvious there are large areas where timber is not precluded.

FIRE RETARDANT TREATMENTS

Certain 'fire proofing' processes are available to resist ignition and flame spread and reduce smouldering. These do not make timber incombustible, but they are effective in suppressing flaming and consequently in cutting down flame spread. On lining materials these could have the effect of converting a Class 3 material into a Class 1 material. There is, however, a difference of opinion as to how effective these retardants are in the case of structural timbers and insurance companies tend to discount them.

There are two methods of application:

(a) Impregnation with chemical salts, either by pressure or hot and cold soaking. (Pressure impregnation is to be preferred, but is most expensive).
(b) Surface coatings.

Chemical salts used for impregnation either give off non-inflammable gases on heating which protect the timber, or melt to form a glaze impervious to the necessary ignition air, or encourage the rapid formation of a heavy layer of charcoal, protecting the layers of timber beneath. Treatment must, however, not be corrosive to metal fixing devices used in the timber assembly, and must not cause hygroscopicity in the wood; that is, at relatively low humidities, cause the timber to become damp.

The best retardants in respect of corrosion and hygroscopicity are monammonium phosphate and diammonium phosphate. Ammonium chloride is also good, but only in mixture with other salts. It must be remembered that most of these salts are soluble in water and therefore must not be exposed to the effects of weather.

Surface coatings are less effective than impregnation because of the thinness of the normal coatings and their vulnerability to damage. They must have good adherence to the wood, must be chemically stable, resistant to high relative humidities and must have an acceptable appearance depending on the proposed location. Most proprietary surface coatings depend on intumescence for their effectiveness (that is they froth on being heated, the froth then hardens and protects the wood beneath). Most common chemical ingredients are sodium silicate, potassium silicate or calcium silicate.

SUMMARY

Timber, although combustible, is still a valid material for the construction of many and various building types, and is accepted as such by the Building Regulations. Its quality of reliability in fire is its greatest benefit and one which many non-combustible materials do not have. Timber is free from sudden changes of state during heating, does not lose strength except through the calculatable erosion of the burning. Timber is a good insulating medium and therefore does not transmit heat which in turn causes the transmission of fire. It does not expand markedly and therefore does not cause peripheral collapse from physical distortion.

12 Industrialised building and timber

First of all it is essential that there is no confusion over the meaning of the terms used in this Chapter. 'Industrialised', 'System' and 'Prefabrication' are terms often used as though they were synonomous. This is not strictly true.

Industrialised building is used in this chapter as meaning building to which the principles of industrialisation have been applied. The RIBA report 'The Industrialisation of Building' (April 1965) referred to the Oxford English Dictionary definition of the verb 'to industralize'—'to affect with or devote to industrialization, to occupy or organize industrially'—and then went on to apply this definition to industrialised building:- 'The organization of building industrially—by applying the best methods and techniques to the integrated process of demand and design, manufacture and construction'. It went on to highlight three areas where the effect of this principle would be recognised (a) the use of factory-made components to a greater or lesser extent (b) the greater mechanisation of site processes (c) the rationalisation of site operations.

It will be seen that this is a far more all-embracing term than prefabricated or system building, although both, carried out on site, might demonstrate all the facets of industrialisation.

System building, in this chapter, is taken to mean a method of rationalising the building operation to produce speed and economy using standardisation, prefabrication (to a greater or lesser extent) and dimensional and design disciplines. This is reflected in the whole building operation being orientated to a factory-made kit of parts, or prefabricated sections of building, which are used to make a *whole* building. A system can be a commercial proprietary method of building which employs prefabricated parts which are special to that system. It is thereby a 'closed' system, in that sense of the word. These could be considered as introverted methods of supplying a building need, as opposed to the extrovert 'open' systems at the other end of the spectrum which are designed by applying the principles of dimensional coordination based on BS preferred range of sizes, and whose parts need not be special to that particular system. A swopping of parts theoretically could take place between systems. Between these extremes there is a wealth of variation. The one incontrovertable fact is that all systems employ prefabricated parts.

Prefabrication is used to mean the pre-forming of parts of a building (or even whole buildings) at some place (usually a factory) remote from site, then transporting them to site and installing them there. The National Building Study—Special Report No 36 ('Prefabrication—A History of its Development in Great Britain') sums up the aim of prefabrication as having been 'the provision of a greater number of building units at greater speed, with the use of less skilled labour (at least

of site labour) and, if possible, at lower cost than could be achieved by
traditional ways of building....'

Prefabrication has been neatly defined by Howard T. Fisher ('Prefabrication. What does it mean to the Architect' Journal of the American
Institute of Architects—November 1948). 'To over-simplify and look a
bit into the future, if you shove and snap a product into place in the
field, that is prefabrication. If you mix, cut, spread, fit and patch, that's
not prefabrication'. This description highlights the differences that
have occured in the building operation in the years since the Second
World War. There are hardly any buildings which are ever constructed
in this country today which do not employ to some degree prefabricated
parts—curtain walling (even standard windows), prehung internal door
sets, staircases, cupboard fittings: to cite just a few examples. And
this highlights the definition predicament.

It could be argued that the concrete block is an example in small scale
of prefabrication. After all is said and done, it is only a difference in
size between the concrete block and the precast storey height wall
panel used by many proprietary concrete building systems. This is
obviously begging the question. After the concrete block, what about
the brick?

Prefabrication in this chapter is taken to imply a definite assembly
process—the forming of parts together into an assembly which is then
transported to site for final installation. This process is becoming
more and more a normal part of building today. The need is to remove
as many operations as possible from the site and into the controlled
conditions of a factory. Advantages gained thereby are more quality
control, more speedy site operations, fewer skilled craftsmen having
to put up with the primitive conditions of a building site and lowering of
costs due to the advantages of mass production.

Therefore to sum up, the whole of the building operation is today moving further and further towards becoming a site assembly process of
as many pre-finished components as possible. Industrialised building
is in the vanguard of this movement, employing factory-made parts to
make whole buildings complete in all respects, and organising the site
processes with the same industrialised principles in mind. System
building, on the other hand, is the actual manufacture of the parts of
a series of buildings governed by a standard design and discipline.
This is prefabrication applied to the whole building, either in parts, or
in sections, and implies a degree of flexibility and design choice in
applying the system to a building problem. Whole buildings can be
prefabricated without flexibility of design choice, and this is pure prefabrication and not system building (cf. the Aluminium bungalows of
the post war period). Nevertheless, as explained above, the whole
building operation today depends on prefabrication to a greater or
lesser extent—but industrialised and system building depend wholly on
prefabrication.

PREFABRICATION IN BUILDING HISTORY

Prefabrication is not a new technique in building. There is evidence that the 'cruck' cottage (referred to in Chapter 2) on occasions demonstrated a very early and primitive form of prefabrication. The main support timbers, or 'crucks' were sometimes cut from the tree, formed in pairs, and then dismantled, numbered and transported to the building site where they were re-erected. Purbeck marble dressings in the thirteenth and fourteenth centuries were often pre-formed in the quarry before transportation to the building they were to adorn.

These examples are, of course, bespoke prefabrication, as opposed to mass produced prefabrication. Mass production had to await the industrial explosion of the nineteenth century, when factory techniques began to be applied to sections of buildings or even whole buildings. We have already seen the outburst in prefabricated timber buildings in America that followed the introduction of the balloon frame. There was a similar outburst in Britain, where there was an enormous increase in production of iron. With the development of cast iron columns and beams repetitive building elements began to be used—and not only elements, but repetitive buildings as well.

One of the first cast iron houses was a canal lock-keeper's cottage at Tipton Green, which dates from 1830. This had walls of flanged cast iron plates bolted together, finished internally with lathe and plaster linings and painted externally.

In 1844 a London firm was shipping warehouses and dwellings to Africa and the West Indies—even a church to Jamaica. The 1849 Californian Gold Rush and the emigration to Australia boosted the market for exported prefabricated buildings. A firm called E.T. Bellhouse of Manchester exported iron houses to San Francisco (and also to Victoria for the gold rush there) and undertook to get the complete shell of a building with ornamented facade and the principle framework of the interior on board a ship within thirty days of the date of the order.

Hemming of Bristol exported iron buildings to Victoria, including galvanised iron churches. The Crimean War, as well, attracted manufacturer's attention and a prefabricated hospital was designed by Isambard Kingdom Brunel, manufactured and shipped to the Crimea.

Nevertheless, of all the examples of prefabrication in Britain during the nineteenth century, the supreme example of repetitive element prefabrication and in many respects the most significant and far reaching in its influence is that of Joseph Paxton's Crystal Palace in 1851. We shall dwell on this example in some detail because in many ways the building established principles of prefabrication applied to a system building which were far in advance of its time, and were not pursued and developed for another hundred years.

Joseph Paxton had started life as a gardener, first in the Horticultural Society's gardens at Chiswick, later as head gardener for the Duke of Devonshire at Chatsworth. It very soon became obvious that he was a

man of exceptional abilities and before long was acting as estates
manager for the Duke, as well as constructing greenhouses, and the
model village of Edensor at the gates of the Chatsworth grounds. He
even at one time ran a newspaper. In 1850 he designed the lily house
at Chatsworth to accommodate a large British Guiana Lily which he
had had imported. Its skeletal structure is said to have derived from
the structure of this lily's enormous leaves. This lily house was the
prototype for the Crystal Palace.

Paxton's design for the exhibition building was produced in 9 days and
ousted the design which had already been approved by the Building
Committee, two members of which were Charles Barry and Brunel.
This was achieved not without difficulty; for the Building Committee
had a proprietary interest in their approved design. It had been, in
fact, submitted by themselves, and had been designed by Brunel. Paxton's design was successful, because it was much more applicable to
the requirements of a temporary exhibition building, than was Brunel's
design with its heavy construction, large sheet iron dome and walls
made up of 15,000,000 bricks.

*Fig.12.1 Crystal Palace, designed by Paxton and built to
house the Great Exhibition of 1851*

Not only would Brunel's building have been impossible to construct in
the time available, it would have cost almost as much to demolish after
the exhibition as to construct. Paxton's design (Fig.12.1) had the
advantage that it was an inspired assembly of light standard parts
(cast iron columns, cross ribbed girders, metal glazing bars and panes
of glass) quick to assemble, making use of techniques of prefabrication
and cheap (1d per cu ft). It was an astounding achievement when it is
considered that the design was carried out in new materials (large
sheets of glass and thin metal sections) without the aid of engineering
skill, merely by the use of rules of thumb and instinct. In fact, the
building took twenty-two weeks to erect, was used for the exhibition
period, was then dismantled and re-erected on a site on Sydenham Hill
in an enlarged form and survived in all for eighty-five years. The
parts of the building were almost entirely re-usable on reassembly.

This building was maybe the first truly industrialised building. Not only was the fullest use made of factory-made standardised assemblies, but the site techniques were integrated to the design, special plant being used to achieve the most economic use of site labour with a minimum of site time. An example of this is the glazing wagon—a staging on wheels designed to run on the gutters on each side of the pitched sections of roof glazing and from which the roof glazing was installed.

Compared with all examples of prefabrication before and most other examples for the next hundred years, the Crystal Palace stands out as a model of systemised thought as opposed to mere prefabrication. The only other examples which come near to this are some of the larger engineering structures of the Victorian period (such as Kings Cross Station, London) where repetitive units were used.

Also in the Crystal Palace, we see the appropriate use of light materials and dry methods of construction to achieve a quick enclosure of space. In this respect, assuming a building of the right scale, timber is an ideal material. However at this period timber was hardly entertained as viable structural material for many building elements, in spite of most roof structures in Britain still being made of wood.

Prefabrication made little progress after 1850 in Britain until the demand for housing after the 1914-18 War encouraged manufacturers to investigate new methods of satisfying this need. Materials used with varying success were pre-cast and insitu concrete (Waller system, Boot House, Easiform House and No-fines Concrete House), steel (Weir House, Dorlonco House and Telford House) and timber (various Scandinavian systems, including the most successful Scano of Scanhouse Ltd).

In 1920 the Committee for Standardisation and New Methods of Construction investigated timber framed houses and recommended the use of the balloon frame with external cladding of weather boarding or rendering on expanded metal as being appropriate to the contemporary need. The most successful timber system used at this time was Scano (Fig. 12. 2), whose wall panels were shipped from Scandinavia. Its costs were however inflated by customs duty, and occasionally labour troubles developed around the system's use. The LCC in the early 1920's experimented with 464 timber framed houses based on the balloon frame technique on the Watling Estate. These were still in good condition in 1958.

The post war boom in prefabrication petered out by 1928 owing to the high cost of houses produced by non-traditional methods as opposed to those build by traditional means. There was little or no experimentation in building types other than housing. A notable exception came in 1936 when C. G. Stillman, Architect for the West Sussex C. C. produced the first schools in Britain in non-traditional construction. These were based on a light steel frame with lattice beams, boarded roof and walls of rendering on expanded metal. Between 1937 and 1940

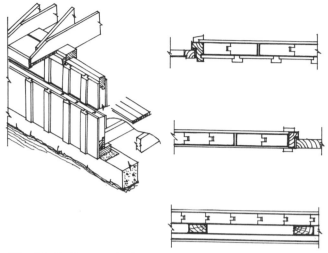

Fig.12.2 Scano timber system. Crown Copyright.

whole schools were built in this method at Selsey, Rustington, North Lancing and Littlehampton. This was the precursor of the Hertford-shire schools after the Second World War and the local authority consortia systems.

Since the Second World War the industrialised building scene has fluctuated from boom to recession, but it is now quite clear that the prin-

Fig.12.3 Aluminium 'prefab'. Copyright: Aluminium Federation.

Fig. 12.4 Aluminium bungalow in four sections being transported to site. Copyright: Aluminium Federation.

ciples of industrialisation are here to stay and prefabrication (as pointed out earlier) is now a growing part of so-called traditional building.

After the War the need to provide immediate housing, and a large aircraft industry with no aircraft to make, led to the manufacture of post war 'prefabs', such as the Aluminium Bungalow (Fig. 12.3). These were made in four sections, completely equipped with kitchen and bathroom fittings and were transported to site on special trailers (Fig. 12.4). Five factories were devoted to the assembly line production of

Fig. 12.5 Assembly line production of aluminium bungalows. Copyright: Aluminium Federation.

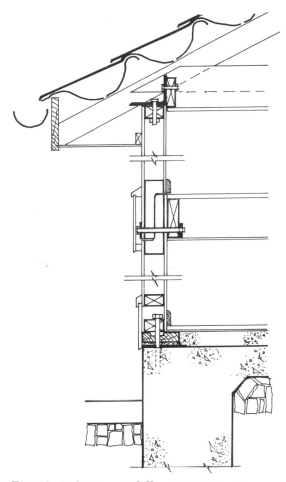

Fig. 12. 6 Scottswood House system. Crown Copyright.

these bungalows (Fig. 12. 5). This was probably the first time that mass
production techniques and flow line methods were applied to whole
building production. The walls were of aluminium with air-entrained
cement grout fill. Whole building sections were handled off the transport with collapsible gantries and slid on trolleys into place on simple
foundations.

Other housing systems to appear included the Scottswood House which
was one of the few timber systems to be marketed during the time of
post war timber shortage (Fig. 12. 6). It consisted of load bearing
timber wall panels with 9 mm plywood cladding both inside and out.
The LCC erected 300 of these houses, but there was considerable concern about the possibility of fire connected with their use.

Steel housing systems to reappear from the between wars period were

the Weir and Atholl houses which had had limited success after the
First World War and a new system—the British Iron and Steel Federation house designed by Frederick Gibberd—which had a modest success.

Concrete systems included pre-cast panel methods such as the New
Duo-Slab System, Airey House, the Wates House, Reema System and
the Dorran House, and insitu methods such as Easiform (Laing) and the
Wimpey House—both of which were no-fines concrete buildings in which
the industrialisation was primarily realised in the rationalisation of
site methods.

Nevertheless, in the vagaries of the system building field since the
war, the most consistently successful housing systems were those
based on no-fines concrete—in fact, the most traditional of the systems
available.

As timber became more readily available after 1952 there was a
flurry of timber housing systems and some manufacturers of steel
houses, such as Weir, turned to timber because of the high cost of sheet
steel. New systems included the Spooner system, Vic Hallam Mk III,
the Calder house and many others which appeared with varying degrees
of success. The Forestry Commission imported Swedish timber buildings for use on their estates; while some UK manufacturers began to
dabble in the timber building export market, such as the Riley Newsum
house which had a limited success in Australia and Canada.

Meanwhile in other types of building, industrialisation was taking great
strides. In education building, C. G. Stillman moved from West Sussex
to Middlesex and the Middlesex school resulted. This was based on the
West Sussex school using a light steel frame. The most sophisticated
developments were, however, happening in Hertfordshire, where the
County architects were developing the first fully integrated Local
Authority system building. On these foundations was built the later
CLASP system and, after this, all the other Local Authority consortia
systems (Fig. 12.7). These were founded on full dimensional co-ordina-
tion and rigid dimensional disciplines, and backed by assured programmes on which bulk purchase agreements could be made for parts of the
buildings as diverse as wall panels and hat and coat fittings.

The lack of assured continuity of demand has been the chief problem
of the commercial developers of building systems. Without this cer-
tainty, the full cost benefit of mass production cannot be realised. In
addition development costs have to be absorbed over a small number
of buildings, thereby inflating their price. This is a problem that the
Local Authority consortia have overcome.

In the years since the War there have been wild fluctuations in the
intensity of industrialised building in this country. Particularly in the
early 1960's the subject was a highly fashionable one and recorded
systems reached the ridiculous number of approximately four hundred.
Many of these systems were no more than paper systems; and of those
that did succeed in reaching a building site, few remain today.

Fig. 12. 7 Hucknall Police Headquarters: a fully integrated system building. Copyright: CLASP Development Group.

However, the principles of prefabrication are now well established across the whole spectrum of building. As explained earlier, the need to use controlled conditions to manufacture parts of buildings and to move certain operations away from the rough outdoor workshops of a building site is being pressured both by the labour force and the requirement of economy and design. Industrialised building is no longer a totally identifiable section of the industry. It is now more a matter of degree than of totally opposed philosophy.

In addition patterns are emerging in the success of various forms of industrialisation. In low rise housing the no-fines systems and some lightweight timber systems seem to be most successful. In the area of high rise development heavy concrete systems such as precast concrete panel systems seem most relevant. But in many areas of low rise commercial and educational buildings, the lightweight systems are very well established. In this area particularly timber has an enormous potential.

TIMBER TODAY IN SYSTEM BUILDING

The work of the Timber Development Association (later becoming the Timber Research and Development Association) finally broke timber out of the emotive strait-jacket it had been in for many years. Basic research was carried out on timber and its characteristics at the Princes Risborough Laboratories of the Building Research Establishment; while TRADA applied the results of this research to practical design projects. The information so gathered was disseminated and a design service was offered to architects and others involved in the building industry. A more rational approach to timber and fire, rot and infestation has resulted.

Timber has become recognised as an appropriate material for use in the lightweight section of the industrialised building field. The reasons for this are:

(a)　suitability of timber to the techniques of factory prefabrication;
(b)　ease of working, requiring no enormous and expensive plant;
(c)　lightness, making its handling inside and outside the factory easy, it being often merely a matter or either manpower, or the simplest of lifting devices;
(d)　the simplicity and cheapness of jigs required to produce timber assemblies;
(e)　good strength characteristics of timber;
(f)　easy fixing devices both on site and in the factory;
(g)　light foundations resulting from a light superstructure.

These factors all contribute to make timber an ideal and economic material for low rise building. With its use, site labour can be cut to a minimum. There are some interesting facts from North America regarding the reduction in site labour made possible by the use of timber prefabrication.

It is claimed that there can be as much as a fifty per cent reduction in the overall labour content by the use of factory-made frames. Such figures as fifty to sixty man/hours are quoted for the erection and sheathing of an average two storey house shell. The all-in times involved, compared with those for traditional building are 500 to 1000 man/hours plus off site labour of 250 to 350 man/hours for non-traditional, compared with 1500 to 2000 man/hours for traditional. Fig. 12. 8 gives an example of a work schedule for the erection of an average single storey timber dwelling shell published by a firm in Connecticut.

Costs can compare favourably with heavier, slower techniques, and where continuity of requirement is assured costs can become very favourable.

PRINCIPLES OF STRUCTURE

All methods of building using prefabricated elements fall into three major categories:

> frame and non-load bearing infill;
> load bearing panels;
> sectional or unitary construction.

Each category can be produced in heavy or light assemblies and can be used for different scales of building.

WORK SCHEDULE FOR AMERICAN HOUSING SYSTEM

Objective: To produce interior and sheathed exterior panels for a house in one day.
All components can be lifted by two men.

TIME	MAN 1.	MAN 2.	MAN 3.	MAN 4.	MAN 5.
8.00	ERECT EXT. WALL PANELS. NAIL SHEATHING NAIL PAPER AT CILL	AS MAN 1.	AS MAN 1.	AS MAN 1.	AS MAN 1.
8.15					
8.30					
8.45					
9.00	PLUMB & BRACE WALLS	AS MAN 1.	FIT TOP PLATE	AS MAN 3.	AS MAN 3.
9.15	ERECT GABLES AND TRUSSES	AS MAN 1.	AS MAN 1.	AS MAN 1.	AS MAN 1.
9.30					
9.45					
10.00	FRAME FOR CHIMNEY	INSTALL SOFFITS FASCIAS ETC.	AS MAN 2.	BRACE GABLE ENDS	AS MAN 4.
10.15					
10.30	LAY CATWALK			LAY AND TACK ROOF SHEATHING	AS MAN 4.
10.45					
11.00	STAIR WELL				
11.15	INSTALL PREFAB. CHIMNEY				
11.30					
11.45	REMOVE BRACES				
12.00					
12.15	LUNCH	LUNCH	LUNCH	LUNCH	LUNCH
12.30					
12.45	ERECT INTERIOR PARTITIONS	CONTINUES SOFFITS, FASCIAS, ETC.	AS MAN 2.	AS MAN 1.	STAPLE ROOF SHEATHING
13.00					
13.15					
13.30					
13.45					
14.00	CEILING BACKERS ETC.			AS MAN 1.	LAY & NAIL ROOFING PAPER
14.15					
14.30					
14.45	PLUMBING BACKERS			CUT OUT DOOR CILLS	
15.00				FRAME BATH TUB	
15.15					
15.30	CLOSET DOOR, FRAMES, ETC.				
15.45					
16.00	REST & DELAY	AS MAN 1.	AS MAN 1.	AS MAN 1.	AS MAN 1.
16.15					
16.30	CLEAN UP	AS MAN 1.	AS MAN 1.	AS MAN 1.	AS MAN 1.

1. MAN 1 IS CREW LEADER AND CARPENTER: MAN 2 IS SECOND IN COMMAND.
2. HATCHED AREAS UNSCHEDULED TIME SET ASIDE FOR UNFORESEEN PROBLEMS.
3. REST + DELAY TIME SPREAD THROUGHOUT DAY.

Fig. 12.8 A work schedule for the erection of an average single storey timber dwelling shell.

Frame and infill

This category refers to any method which is based on a structural frame, whether of timber, steel or concrete, clad or infilled with non-

load bearing wall panels. This principle is illustrated in Fig. 12. 9.
(The Crystal Palace was the first example of this type of system
building.)

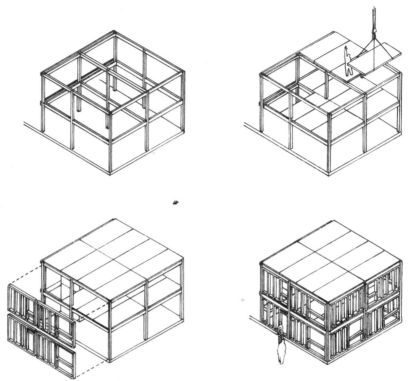

Fig. 12. 9 System building based on a structural frame.

One distinction must be made immediately; although this category is
designed on frame-action principles, these principles are not exclusive
to this category. Sectional or unitary constructions can also be designed
using frame-action principles.

Many of the steel houses referred to earlier were designed on this
frame and infill principle. The Weir and Atholl houses are such examples. Often the frame was structurally almost superfluous as the walls
were strong enough to do the work by themselves—hence a lack of
economy. The frame and infill principle works best for non-domestic
buildings, and the CLASP and OXFORD systems based on a light steel
frame are good examples of the principle being used in its right context (Fig. 12. 10).

Long span industrial systems have successfully sprung up based on
the use of the precast concrete or steel portal frames—and more
recently the steel space frame, which is at the moment restricted in
its application because of its expense.

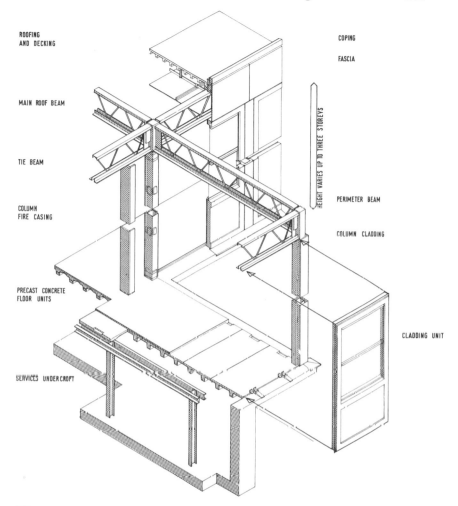

ROOFING
AND DECKING

COPING

FASCIA

MAIN ROOF BEAM

HEIGHT VARIES UP TO THREE STOREYS

TIE BEAM

PERIMETER BEAM

COLUMN
FIRE CASING

COLUMN CLADDING

PRECAST CONCRETE
FLOOR UNITS

CLADDING UNIT

SERVICES UNDER CROFT

Fig. 12.10 Isometric of Oxford System. Copyright: Oxford Regional Hospital Board and C.E.D. Building Services.

Timber has been used widely for frame buildings up to two storeys in height, and we may well see a wider use of the rigid frame principle making a real challenge in some areas of the portal frame industrial field.

The frame and infill category is a kit-of-parts method of building. The frame is designed to a dimensional discipline with standard spans and member sizes and standardised junction details. Infilling would probably be in prefabricated wall panels, maybe with complete external treatments applied in the factory (including glazing and decoration).

Internal linings would be omitted to allow all fixing to be made internally and engineering first-fix services to be installed prior to lining out. This method can avoid entirely the use of scaffolding.

Floor and roof panels are normally prefabricated and may often (in the case of the roof panels) be complete with the first layer of the roof covering. Components are often (in the light systems) designed so that they can be manhandled on site, cutting the use of expensive cranage to a minimum, and to only certain parts of the erection procedure. Other systems (such as Anderson A75 and early CLASP) employ heavy precast concrete cladding systems that rely on constant cranage during the shell erection, but provide maintenance free finishes. These heavier systems are more appropriate to buildings over two storeys in height.

Many multi-storey flat systems employ a frame system. One of the most interesting is the Wimpey 1001 system which uses cast insitu concrete—dense concrete for the structural frame and no-fines concrete for the wall infills. This method relies more on the standardisation of shutters rather than the prefabrication of the components. But the total conception is fully industrialised. Design is integrated with site operations. Shutter boards are re-used continuously on a sequence basis with window and door assemblies fixed inside the shutters prior to pouring the no-fines concrete. Windows can be pre-glazed and decorated before fixing inside the shutters.

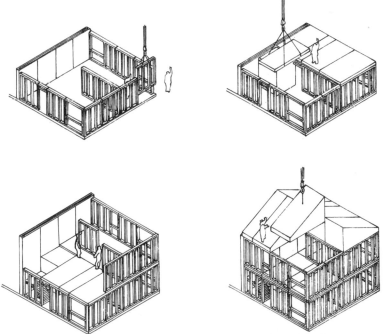

Fig. 12.11 Load bearing panels made up of timber framing.

Load bearing panels

This category covers those systems which consist of an assembly of
load bearing wall panels between which the beams span and which con-
vey those loads to the foundations without the use of independent
columns. The structural principle is therefore not frame action, but
post and lintel and bracing panels. The form the panels take can be
very varied, three of which as illustrated. They are all, however, prefabricated and governed by a dimensional discipline. Fig. 12. 11 shows
panels made up of timber framing, Fig. 12. 12 precast panels of lightweight concrete and Fig. 12. 13 precast heavy concrete. There have
even been load-bearing panels systems of cast iron, an example of this
has already been referred to at Tipton Green dating from 1830.

The lightweight systems are more appropriate to low rise buildings
than the heavy systems. There are many heavy concrete panel systems
used for high rise flats. The Ronan Point flats were carried out in one
of these, and in spite of the set back that these systems have experienced after the disaster there, similar methods are appropriate to this
type of development. Often in these types of buildings, the contractor,

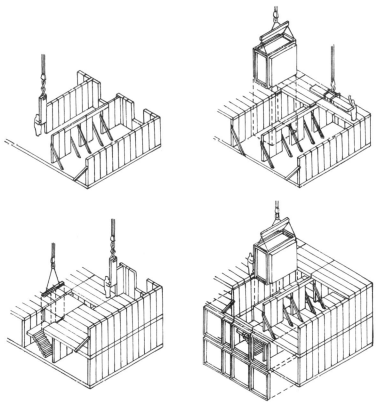

Fig. 12.12 Precast panels of lightweight concrete.

to avoid the heavy transport cost of moving the precast units from the factory to the site, sets up a site factory for casting the units. The contract, however, needs to be large for this to be economic.

This method of building is again a kit-of-parts method and many of the comments on the frame and infill method are applicable to this load-bearing principle.

Timber is very appropriate for this category of building up to two storeys. Case Study No 2 which follows will deal in detail with such a system.

Sectional and unitary construction

This category covers any system that consists of constructing a section of the finished building in a factory, to a greater or lesser extent complete with all finishes, and then transporting it to site where the whole building is completed by attaching the sections together.

This is the category that includes the greatest element of prefabrication and the greatest need for standardisation. One of the first examples of this carried to a logical conclusion was the aluminium bungalow

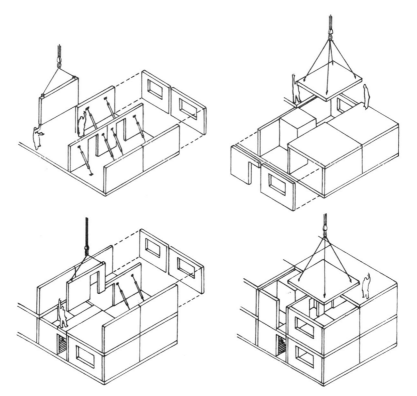

Fig. 12.13 Precast panels of heavy concrete.

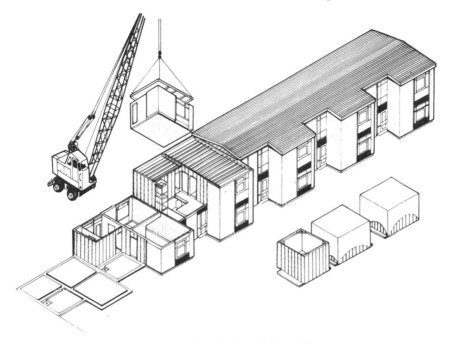

Fig. 12.14 Locarn system using lightweight ribbed concrete panels. Copyright: Foster Construction Services.

Fig. 12.15 Atrium house, a Danish example of a unitary building system. Copyright: A.S. Byggeselskabet.

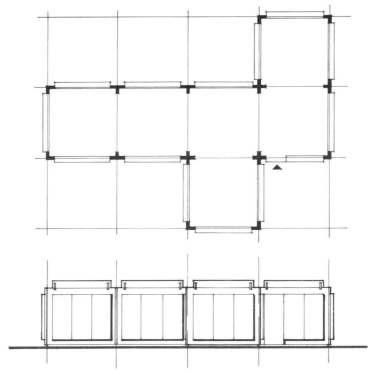

Fig.12.16   Arne Jacobsen's Kubeflux system.   A.Jacobsen.

after the Second World War. Other examples include the relocatable
timber buildings often associated with temporary school buildings,
since the War. These can be in the form of skid units—sections of
building totally finished in the factory, pushed into position on site and
bolted one to another, or in a folded or pack form such as the various
buildings produced for this market by Terrapin International Limited.
Two storey houses, marketed by Calders Limited, were made up of four
box units of timber framing, clad internally and externally and completely fitted out with services. These were handled on site by a crane
and it was claimed that site operations were cut to a minimum so that
a mere five hours would suffice to erect a terrace house. Timber in
these examples is obviously the ideal material because of its lightness,
but there have been attempts to use other materials, as in the Locarn
System that used lightweight ribbed concrete panels in an interesting
compromise between this method and the load-bearing panel method.
In this case the panels were prefabricated in the factory and transported to site, where they were assembled into room size sections of
the building and handled into the building position by crane (Fig. 12. 14).

Nevertheless the requirement for lightness is obviously essential,
otherwise the handling problem becomes impossible. The one major
disadvantage of this type of method is that being so highly industrialised and being capable of being totally prefinished in the factory, there
is a limitation of choice, both in the actual design and layout of buildings so realised and the finishes they exhibit.

Two interesting Danish examples of unitary building systems are the
Atrium House (Fig. 12. 15) where the house shell only is prefabricated;
and Architect Arne Jacobsen's Kubeflex system (Fig. 12. 16) based on a
cube shaped unit of laminated wood frames with timber roof and floor
panels—the whole building being of wood and providing within its repetitive form a considerable measure of planning flexibility.

In the two chapters that follow, we shall take two case studies—one of a
unitary system, one of a load bearing panel system—both using timber
as the appropriate and economic answer to a particular application;
illustrating the use of timber in an area where its properties are put
to the greatest use—the prefabrication of low rise structures.

13 Case Study 1—Unit system

This and the following chapter illustrate the use of two timber based systems to solve accommodation problems in the educational field requiring minimum construction times.

PROBLEM

Because of an unexpected influx of children of infant school age into an area, due to the building of several housing sites, a local education authority was faced with an urgent requirement for additional classrooms in an already overcrowded local school. A new infants school was planned for the area, associated with a new local authority housing scheme; but this project was not due to commence for two years. As a result it was expected that the emergency requirement for extra classroom space was only temporary.

SOLUTION

It was decided to use a lightweight unitary timber system building to overcome the problem.

This building system has the advantage of easy relocatability, being a modular system made up of buildings (or units) of 20 m² floor area. Each unit travels to site on a transporter and is then handled into position by crane as a finished section of building.

The floor is of redwood joists between steel Z beams with a glued and nailed chipboard deck acting as a stressed skin panel. A Lino tile floor finish is installed in the factory. The roof is made up of plywood stressed skin decking and bituminous felt roof covering and is supported by timber columns. The ceiling is of plasterboard already decorated in the factory and polystyrene is installed in the cavities in the roof frames. The wall panels are plywood faced externally and finished with a rough textured plastic composition, and lined internally with plasterboard. Window units are glazed and decorated in the factory.

Each unit is positioned immediately adjacent to its neighbouring unit and instant weather protection is provide by a metal flashing between roofs and coverboards internally and externally between wall panels.

The advantages of this type of timber system to solve this particular problem are:

>Speedy erection due to the large pre-finished elements of building.
>It is possible to produce and transport the units economically, because of the lightness of timber.
>Simple foundations due to the lightness of the structure.
>Easy removability of the building when the need for it passes.
>The almost 100 per cent re-usability of the units of building on another site either in the same form or some other plan form.

STRUCTURAL ASPECTS

The component system described in Chapter 14 is essentially a kit of parts, factory made and transported to the site. Lateral stability achieved by the introduction of bracing panels on the perimeter and in the case of buildings in excess of about 15 m in length, internal bracing panels may also be necessary. The use of internal bracing panels reduces the planning flexibility of the component system but on the other hand the number of parts in the kit can be increased to give a larger range of applications, for example longer clear floor and roof spans. However, the fewer the number of parts the greater the possibility of achieving flow line production and this is a vital economic consideration.

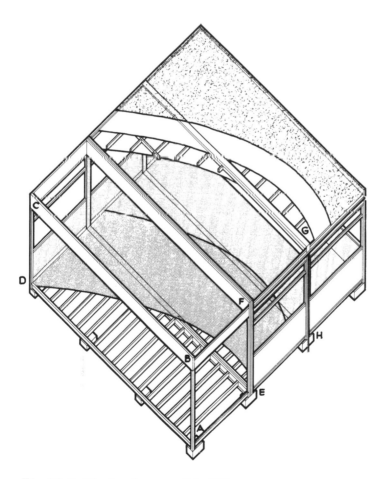

Fig. 13.1 Single storey unit building

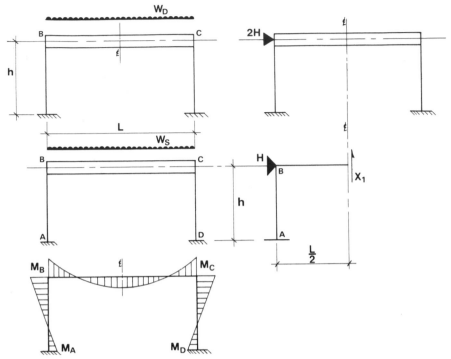

Fig. 13.2   The analysis of a typical frame

With the unit system a number of structural elements are assembled
to form a complete section of building which travels to the site in section form as described previously. The erected sections butt against
each other and thus are suited to the requirement of providing additional classrooms, three units giving a plan area of 60 m². Each unit is
structurally independent and relies on frame action to achieve lateral
stability (Fig. 13.1). The capacity of the structure is limited by the
size of the timber columns which are 75 mm × 75 mm in cross section.
For exposed sites subjected to high wind speeds the timber columns
are replaced by rectangular hollow section steel columns. The analysis of a typical frame ABCD Fig. 13.1 is outlined below. The method
of assembly of each unit is such that the roof beam BC acts as a simply
supported member for dead load but frame action is assumed for imposed (snow) and wind loading. If the beam BC carried a dead loading
Wd per unit length, the maximum bending moment, Fig. 13.2, is given by

$$Md = \frac{WdL^2}{8}$$
13(i)

For the imposed loading Ws per unit length the bending moments at

B and C, Fig. 13.2, are given by

$$M_B = M_C = K_1 \times \frac{WsL^2}{6}$$ 13(ii)

where $K_1 = \dfrac{\dfrac{Ic}{h}}{\dfrac{2Ic}{h} + \dfrac{Ib}{L}}$

Where Ic is the second moment of area column, Ib is the second moment of area of beam. The maximum span moment is given by

$$Ms = \frac{WsL^2}{8} - M_B$$ 13(iii)

With regard to lateral loading (wind) the structure can be made statically determinate by making a cut at the centre of the beam BC, Fig. 13.2, and due to symmetry the only unknown is the force x_1. If the lateral load is 2H then x_1 is given by

$$x_1 = \frac{h^2H}{L\left[h + \dfrac{L}{6K_2}\right]}$$ 13(iv)

where $k_2 = \dfrac{Ib}{Ic}$

For lateral loading in the plane EFGH or similar procedure can be adopted in which EFGH is treated as a rigid frame.

Some typical numerical results are evaluated below where L = 7.2 m, h = 2.68 m, Ws = 1.0 KN/m and Wd = 0.65 KN/m.

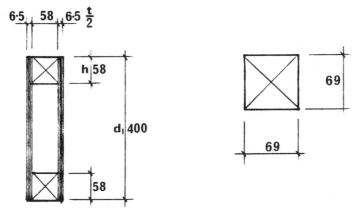

Fig. 13.3 Beam and column sections

The beam and column sections are shown in Fig. 13.3 giving $Ib = 2.68 \times 10^8 mm^4$ and $Ic = 1.89 \times 10^6 mm^4$

To determine the factor K_1

$$\frac{Ic}{h} = \frac{1.89 \times 10^6}{2.68 \times 10^3}$$

$$= 705$$

$$\frac{21c}{h} = 1410$$

$$\frac{Ib}{L} = \frac{2.68 \times 10^8}{7.2 \times 10^3}$$

$$= 37222$$

Thus $K_1 = \dfrac{705}{1410 + 37222}$

$$= 0.018$$

Thus from equation 13(ii)

$$M_B = M_C = k_1 \times \frac{WsL^2}{6}$$

$$= \frac{0.018}{6} \times 7.2^2 \times Ws$$

$$= 0.158 \, Ws$$

Thus for $Ws = 1.0$ KN m

$$M_B = M_C = 0.158 \text{ KN m}$$

The midspan moment is obtained from equation 13(iii)

$$Ms = 1.0 \times \frac{7.2^2}{8} - 0.158$$

$$= 6.48 - 0.158$$

$$= 6.322 \text{ KN m}$$

The midspan moment due to dead loading is obtained from equation 13(i)

$$Md = 0.65 \times \frac{7.2^2}{8}$$

$$= 4.2 \text{ KN m}$$

Thus maximum beam moment Mt

$$= Md + Ms$$

$$= 6.322 + 4.22$$

$$= 10.522 \text{ KN m}$$

The moment at midspan of the beam due to lateral loading is zero. For the beam section shown in Fig. 13.3 the maximum chord stress is given by

$$f = \frac{Mt \times y}{Ib}$$

$$= \frac{10.522 \times 10^6 \times 200}{2.68 \times 10^8}$$

$$= 7.92 \text{ N/mm}^2$$

The maximum shear force is given by

$$Q = (Wd + Ws) \frac{L}{2}$$

$$= 1.65 \times 3.6$$

$$= 5.94 \text{ KN}$$

The shear stress at the ply web, softwood chord interface is given by

$$qyy = \frac{QA\overline{Y}}{2hIb} \text{ (see equation 9(xvi)}$$

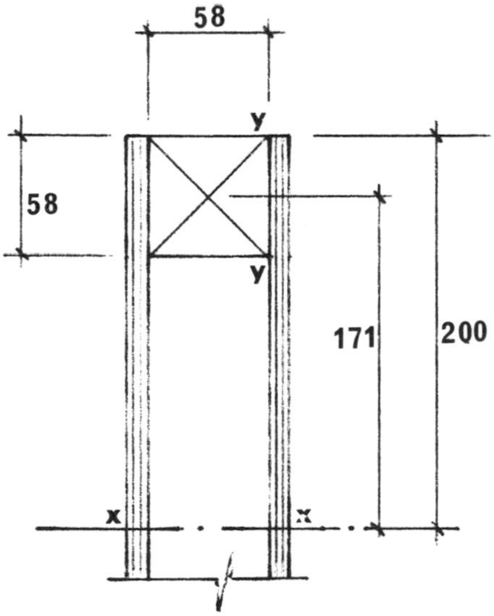

Fig. 13.4 Notation and dimensions for calculating shear stress.

Using the notation in Fig. 13. 4

$$qyy = \frac{5.94 \times 10^3 \times 58 \times 58 \times 171}{2 \times 58 \times 2.68 \times 10^8}$$

$$= 0.11 \text{ n/mm}^2$$

The panel shear stress is obtained from equation 9(xiv)

$$qp = \frac{QAY}{Ibt}$$

where $Ay = \dfrac{b_2 h}{2}(d_1 - h) + \dfrac{td_1^2}{8}$

$$= \frac{58}{2} \times 58(400 - 58) + \frac{13 \times 400^2}{8}$$

$$= 575000 + 260000$$

$$= 835000$$

thus $qp = \dfrac{5.94 \times 10^3 \times 0.835 \times 10^6}{2.68 \times 10^8 \times 13}$

$$= 1.42 \text{ N/mm}^2$$

For a lateral load of say 2H = 950 N then x_1 can be evaluated from equation 13(iv)

$$K_2 \quad \frac{Ib}{Ic} = 142$$

$$x_1 = \frac{2.68^2 \times 475}{7.2 \left[2.68 + \dfrac{7.2}{6 \times 142} \right]}$$

$$= 176 \text{ N}$$

Thus $M_B = 176 \times \dfrac{L}{2}$

$$= 176 \times 3.6$$

$$= 633 \text{ Nm}.$$

The column section is checked for the worst combination of direct loading and bending moment in accordance with the procedure outlined in Chapter 9. For the design of the columns a duration of load factor Kd = 1. 5 (see Chapter 8) is appropriate for the combination of snow and wind loading.

14 Case Study 2—Component System

PROBLEM

There was an urgent requirement for a building to re-house a university department due to the unexpected termination of the lease of an old building in which the department was housed. The only available site was in the grounds of a hall of residence at some distance from other academic buildings of the university. It was decided to build a structure that could ultimately revert to social use associated with the residential accommodation when the department moved to the new academic block eventually to be part of the new university campus.

Permanence was a requirement of the structure, but flexibility, to allow for large scale modifications caused by the expected change of usage, was essential. Speed of construction was important in order to have the new building fully functional by the beginning of the new academic year.

SOLUTION

It was decided to use a lightweight timber component system building as this would provide the flexibility necessary to cope with the changing levels of the site, speedy erection and future adaptability. In addition the system was capable of satisfying the requirements of engineering performance and permanence.

The system, the elements of which are shown in Fig. 14.1, is based on a load bearing external timber wall panel of European Redwood with brick/plywood claddings. The wall panels are erected on a reinforced concrete ring beam and ground floor slab foundation. Stressed skin plywood box beams (both first floor and roof) span between wall panels, or between wall panels and rectangular hollow section internal columns. Prefabricated floor and roof panels are made up of redwood framing and birch plywood decking, acting as stressed skin panels, the roof panels having the first layer of roofing felt bonded to them in the factory. All window units are glazed and decorated in the factory and are inserted and fixed into apertures in the wall panels. All the external surfaces of the wall panels are totally finished in the factory, to eliminate much of the need for scaffolding and to provide a watertight shell as soon as possible. All components can be handled on site by two men, with the exception of the longer span beams, which require placing by crane. In the case of a two storey building a crane is required on two occasions, once to place the floor beams and once to place the roof beams. At these times all the components for the continued building operation are handled up to the appropriate level by crane (i.e. first floor panels and wall panels at the time of placing the first floor beams).

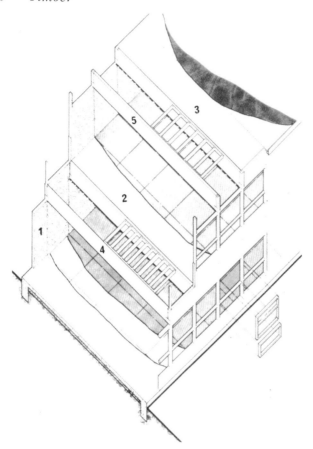

*Fig.14.1 Elements of a lightweight timber component system.
1 wall panel: 2 floor panel: 3 roof panel: 4 floor beam: 5 roof
beam.*

A steel continuity strip is included between wall panels from ground
slab to half way up the first floor wall panels to provide the additional
load bearing capacity in the standard wall panels at the ground floor
level and link ground and first floor panels. Stressed skin bracing
panels are introduced as appropriate to provide the right degree of
resistance to wind pressure depending on the size and shape of the
building and its degree of exposure.

All internal linings are installed with the engineering services in the
totally weather-proof shell at a much earlier date than would be possible if the building had been built traditionally. Internal partitioning
(except where additional bracing partitioning is required) is totally
demountable and therefore the internal division is fully flexible. Much
of the engineering services can be prefabricated due to the 1.800 m
plan discipline.

The building was of a total floor area of 1900 m² on two floors. It was designed on a split level arrangement because of the slope of the site and the substructure work included brick and concrete retaining walls. The total contract period was six months. The whole of the weathertight shell was erected by a four man team in nine weeks.

To summarise, the essential features of the system are as follows:

 adaptability on plan;
 elevational adaptability, including the use of traditional external
 materials such as brickwork, without impeding the building programme;
 lightness, giving speed of construction, good erection economics,
 and simple foundations.

STRUCTURAL ASPECTS

Preliminary investigations indicated that for the wall panel thickness envisaged it would not be possible to design the structure as a series of frames with rigid joints (Fig. 14. 2). Also it was considered to be desirable to simplify as far as possible the connections between the wall panels/columns and the beams with a view to minimising the metalwork and speeding up erection. Thus a post and lintel solution was adopted which requires bracing panels to maintain stability (Fig. 14. 3). For maximum planning flexibility it is preferable to limit the

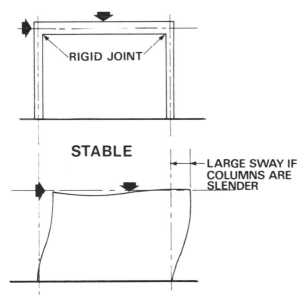

Fig. 14.2 Limitation of frame with rigid joints.

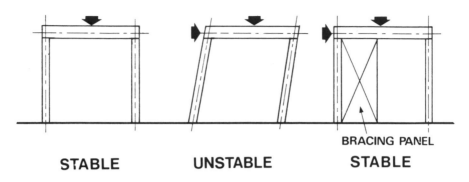

Fig.14.3 Post and lintel.

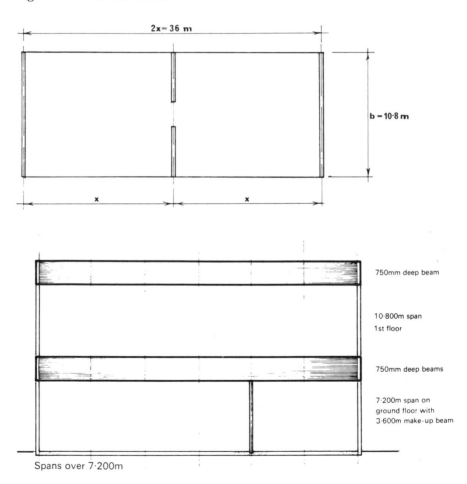

Fig.14.4 Above layout of bracing panels. Below Beam layout.

positions of the bracing panels to the gable ends only. However the layout of this particular structure was such that one the wings of the building had plan dimensions of 36 m × 10.8 m and a line of internal bracing panels was required (Fig. 14.4a). The roof and floor beams were of plywood box section at a spacing of 1.8 m and were designed as simply supported over spans of 10.8 m, 7.2 m and 3.6 m (Fig. 14.4b). The transfer of lateral (wind) loading to the bracing panels is outlined below.

The assessment of wind loading was based on the British Standard Code of Practice CP3: Chapter V: Part 2: 1970. The basic wind speed V, appropriate to the district in which the structure was erected was taken as 45 m/s. The design wind speed Vs was calculated from the expression

$$V_S = VS_1S_2S_3 \qquad\qquad 14(i)$$

where S_1 is a topography factor, taken as 1.0

S_2 is a factor which takes into account the combined effects of ground roughness, the variation in wind speed with height above ground and the size of building of component part under consideration.

S_3 is a statistical factor which takes into account the degree of security required and the design life for the structure. A fifty year design life was assumed with a probability level of 0.63 thus giving a value of unity for S_3.

From the value of the design wind speed V_S obtained from equation 14(i) the dynamic pressure head q given by

$$q = 0.613 \text{ Vs}^2 \qquad\qquad 14(ii)$$

where q is expressed in N/m² V_S in m/s.

From the overall dimensions of the building the lateral force per unit area p may be expressed as

$$p = Cfq \qquad\qquad 14(iii)$$

where Cf is a force coefficient for rectangular clad buildings obtained from table 10, CP3: Chapter V: Part 2.

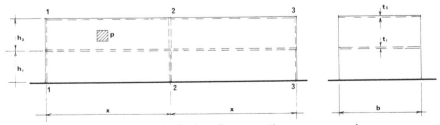

Fig. 14.5 Notation for calculating force actions on one wing of structure.

Consider this force p acting on the largest face 2x (Fig. 14.5) of one wing of the structure. It is assumed that roof and floor panels which are nailed to the box beams act as diaphrams transmitting the loads to the bracing panels along lines 1-1, 2-2 and 3-3. For the half length x the total wind force is given by

$$F = px(h_1 + h_2) \qquad\qquad 14(iv)$$

The force at roof diaphragm level

$$= px\,\frac{h_2}{2} \qquad\qquad 14(iv)$$

The force at intermediate floor level

$$= \frac{px}{2}(h_1 + h_2) \qquad\qquad 14(v)$$

The shear at each end of the roof diaphragm of length x is

$$Q_D^R = \frac{Pxh_2}{4} \qquad\qquad 14(vi)$$

The shear at each end of the floor diaphragm of length x is

$$Q_D^F = \frac{Px}{4}(h_1 + h_2) \qquad\qquad 14(vii)$$

Considering the roof diaphragm to act as a horizontal beam of length x and depth b the maximum bending moment is given by

$$M = \frac{Pxh_2}{2}\cdot\frac{x}{8}$$

$$= \frac{Ph_2 x^2}{16} \qquad\qquad 14(viii)$$

If the roof diaphragm thickness is t_R the stress at A and B, Fig. 14.6,

$$f = \frac{Ph_2 x^2}{16}\cdot\frac{6}{b^2 t_R}$$

$$= \frac{0.375\,Ph_2 x^2}{b^2 t_R} \qquad\qquad 14(ix)$$

The force per unit length at the extreme fibres is equal to $f.t_R.1$

$$= \frac{0.375\,ph_2 x^2}{b^2} \qquad\qquad 14(x)$$

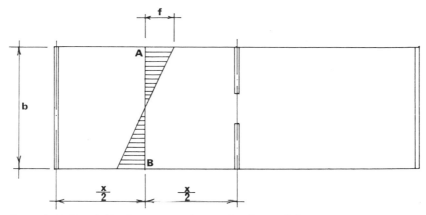

Fig.14.6 Roof diaphragm acting as horizontal beam.

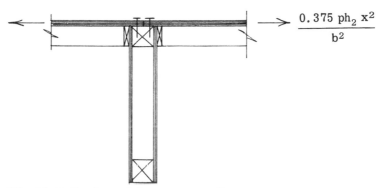

Fig.14.7 Roof panel beam connection.

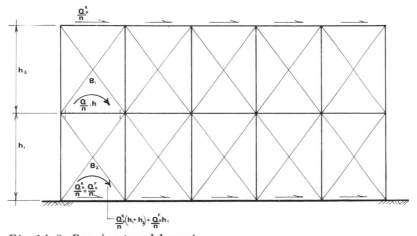

Fig.14.8 Bracing panel layout

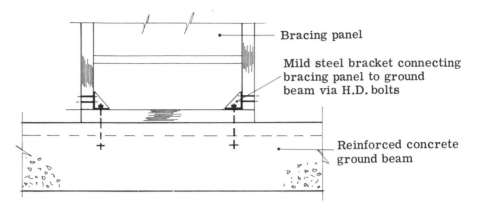

Bracing panel

Mild steel bracket connecting bracing panel to ground beam via H.D. bolts

Reinforced concrete ground beam

Fig. 14.9 Holding down details for bracing panels

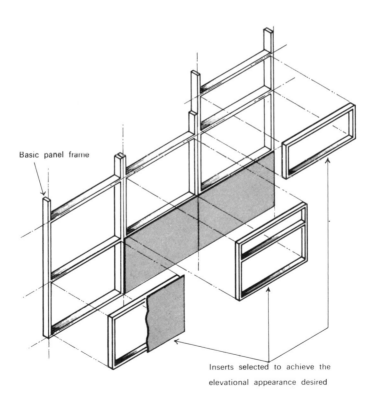

Basic panel frame

Inserts selected to achieve the elevational appearance desired

Fig. 14.10 Standard wall panels.

The roof panels are nailed to the plywood box beams (Fig. 14.7) and thus this force must be resisted by the nails in lateral load if the full diaphragm action is to be realised. If N is the permissible lateral load per nail then the number of nails per unit length required at A and B is

$$= \frac{0.375\, ph_2 x^2}{b^2 N} \qquad\qquad 14(xi)$$

Using similar reasoning the number of nails per unit length required at midspan section of the intermediate floor is

$$= \frac{0.375\, px^2(h_1 + h_2)}{b^2 N} \qquad\qquad 14(xii)$$

Let n equal the number of bracing panels at each end of the roof dia-phragm. The lateral load for bracing panel is $\frac{Q_D^R}{n}$ and thus a typical bracing panel B_1 (Fig. 14.8) is required to resist a bending moment $\frac{Q_D^R h_2}{n}$ and a shear $\frac{Q_D^R}{n}$.

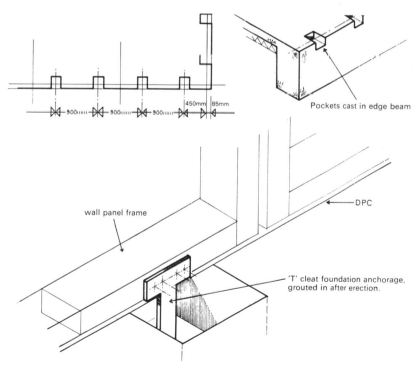

450mm 85mm

—300ıııı— —300ıııı— —300ıııı—

Pockets cast in edge beam

wall panel frame

DPC

'T' cleat foundation anchorage, grouted in after erection.

Fig. 14.11 Fixing details for standard wall panels.

Panel B_1 is bolted to panel B_2 at intermediate floor level and B_2 is designed for a shear of $\dfrac{Q_D^R}{n} + \dfrac{Q_D^F}{n}$ and a bending moment of $\dfrac{Q_D^R}{n} \cdot (h_1 + h_2) + \dfrac{Q_D^F}{n} h_1$. These loads are transmitted to the reinforced concrete ground beam via corner angles and bolts (Fig. 14.9). The floor and roof panels, floor and roof beams were designed by the methods outlined in Chapter nine and particular attention was paid to local wind load effects at the edges and corners of the roof and wall panels.

Fig. 14.12 General view of component system partially erected.

A similar procedure was adopted for estimating the wind loads on the shorter face b, but solid bracing panels were not required as it was assumed that the load could be transmitted to the ground beams via 20 standard wall panels (Fig. 14.10) in the 36 m length. The method of fixing these wall panels to the ground beam is shown in Fig. 14.11. A feature of this component system is the extensive use of nails to provide structural connections. High tensile twist nails were used to connect the floor and roof panels to the beams using a standard nailing pattern based on equations 14(xi) and 14(xii). Fig. 14.12 is a general view of the system, partially erected, showing the continuity strip which connects the ground and first floor wall panels, the floor beams and the floor panels stacked prior to nailing to the top chords of the beams. This was day 10 of the erection of the 36 m $\times$ 10.8 m wing indicating the speed with which the structure can be erected.

15 Testing

In Chapter 5 it was demonstrated that basic stresses for use in the design of timber structures are derived from short term tests on small clear specimens. Tests on built up components, joints and complete structures also form an important aspect of timber engineering. There are a number of situations in which testing is desirable and these are listed below.

As a means of quality control in the factory production of structural components such as box beams or trusses. Fig. 15.1 shows a rig designed for testing box beams forming part of an industrialised building system. The beam is undergoing a deflection and strength test in accordance with the procedure laid down in CP112, which will be described later.

Fig. 15.1 Rig for testing box beams.

There are a large number of fixings used for timber joints which are not covered by Codes of Practice and further, they are not generally amenable to calculation. Fig. 15.2 shows a nailed joint between plywood and softwood, forming a roof diaphragm, under test. The number of tests carried out should be sufficient to permit a statistical analysis to be made.

In the development of timber building systems it is sometimes possible to test a complete structure. The rig shown in Figs. 15.3 and 4 was used in the development of a unit building system to simulate lateral loads due to wind. In this way the overall response of the system to loading, rather than that of the individual components, can be examined.

The method of testing components laid down in CP112 is given below.

PRELOAD. A load equal to the design long term load is applied and maintained for a period of thirty minutes and then released. Where

Fig.15.2 *Nailed joint between plywood and softwood under test.*

Fig.15.3 *Rig used to simulate lateral loads due to wind.*

Fig.15.4 Jacking towers and spreader beam for applying lateral load

camber is provided, this is recorded in relation to a datum through the support points after the load has been released. The deflection of the member is measured before and after the load is released, and again at the end of a further fifteen minutes.

DEFLECTION TEST. Immediately following the preload test, the design long term load is again applied. This is maintained for fifteen minutes and then additional load is added until the maximum design load is attained. Loading is such that the time taken to reach design load from zero load is at least thirty minutes and preferably not more than forty-five minutes. Maximum load is maintained for twenty-four hours and then released. Deflection readings are taken during the test as follows:

fifteen minutes after application of long term load;
fifteen minutes after application of maximum design load;
at sufficient intervals throughout the twenty-four hours under maximum design load to enable a deflection/time curve to be plotted;
at the end of the twenty-four hour period and
fifteen minutes after release of the load.

STRENGTH TEST. After the interval of fifteen minutes, the maximum design load is then reapplied as under the deflection test. The load is increased at the same uniform rate until $2\frac{1}{2}$ times the design load is applied.

The object of the above procedure is to assess the performance of the component in terms of deformation and strength and this will be examined in relation to a series of tests carried out on ply web box beams.

The primary object of these tests was to assess the performance of birch faced spruce core plywood web box beams in terms of ultimate strength and load/deformation characteristics with that of a single species plywood web (birch) beam. A summary of the results is given below:

	Total deflection at design load mm	Load at failure KN	Load factor failure load design load
Beam No. 1 Birch/Spruce webs	45	21	2.06
Beam No. 2 Birch webs	39	30	2.93
Beam No. 3 Birch/Spruce webs	36	30	2.93
Beam No. 4 Birch/Spruce webs	41	28	2.77

The design load for the beams is 10.2 KN and the cross section and loading arrangement are shown in Fig. 15.5. The preload and deflection tests are specified in order to assess the deflection characteristics. It was emphasised in Chapter 8 that under sustained loading the deformation of timber increases with time and it is important to establish that the rate of increase in deflection decreases with time (see curve OA

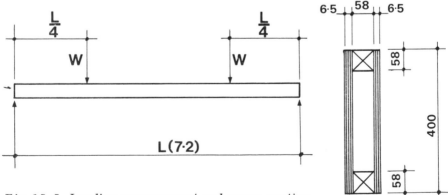

Fig.15.5 Loading arrangement and cross section.

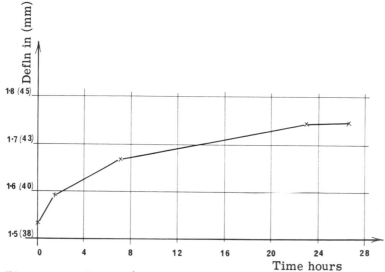

Fig. 15.6 Deflection/time curve.

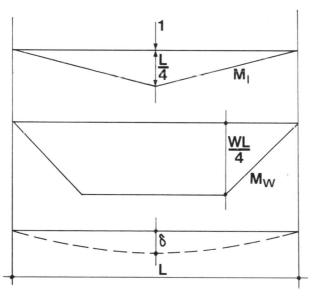

Fig. 15.7 Bending moment diagrams for unit load and two points loads of W.

Fig. 8.1). The deflection/time curve at design load for beam 2 (solid birch web) for a period of twenty-seven hours is shown in Fig. 15.6 which indicates that the rate of increase in deflection is decreasing. For loading up to the design value (apart from initial slip at the bolted end connections) the load/deflection relationship for all the test beams was linear. The midspan deflection, ignoring shear deformation, may be obtained by applying a unit load as in Fig. 15. 7.

$$\delta = \frac{M_1 \, Mwds}{EI}$$

where EI is the flexural rigidity, assumed constant

thus $\delta = \dfrac{2}{EI}\left[\dfrac{1}{3} \times \dfrac{L}{4} \times \dfrac{L}{8}\dfrac{WL}{4} + \dfrac{L}{4} \times \dfrac{L}{6}\left(\dfrac{WL}{4} \times \dfrac{L}{8} + 4 \times \dfrac{WL}{4} \times \dfrac{3L}{16} + \dfrac{WL}{4} \times \dfrac{L}{4}\right)\right]$

$$= \frac{2}{EI}\left[\frac{WL^3}{384} + \frac{9WL^3}{2 \times 384}\right]$$

$$= \frac{0.0286 \, WL^3}{EI} \qquad\qquad\qquad 15(i)$$

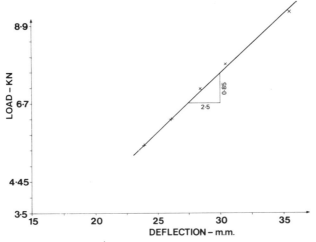

Fig. 15. 8 Load/deflection curve.

Considering the load/deflection curve for beam 2, Fig. 15. 8 it is possible to obtain a value for the apparent modulus of elasticity of the beam if the value of the second moment of area I is obtained from the dimensions shown in Fig. 15. 5. The load/deflection relationship is linear and for a loading increment 2w = 0. 85 KN the deformation is 2. 5 mm. Thus from equation 15(i) the apparent modulus of elasticity can be determined.

$$\delta = \frac{0.0286 \, WL^3}{EI}$$

$$E = \frac{0.0286 \ WL^3}{E\delta}$$

$$I = 2.34 \times 10^8 mm^4$$

$$\text{thus } E = \frac{0.0286 \times 0.425 \times 7.2^3 \times 10^9}{2.34 \times 10^8 \times 2.5}$$

$$= \frac{0.286 \times 4.25 \times 7.2^3}{2.34 \times 2.5}$$

$$= 7755 \ N/mm^2$$

This value of E approaches the mean value for a species group S2 timber (see Chapter 9) and justifies the assumption of using the mean value for calculating the deflection of box beams.

In terms of serviceability the results for beams 2, 3, and 4 appear to be satisfactory. With beam 1 the load at failure was considerably less than for the remaining beams and this was due to the presence of a knot close to a lifting hole. This defect was large enough to warrant rejection of the chord member on the production line.

It is obviously desirable to test more than one component but testing is an expensive operation and therefore it is not possible to test a number of components in the same way that a large number of clear specimens are tested to obtain basic stresses. However the scatter of results is not as wide as for individual specimens of timber. The Code assumes that if five tests are carried out the minimum of the five must achieve a load factor of two times design load. This load factor is increased if the number of tests is less than five in accordance with the table below.

No. of similar components tested	Minimum load factor
1	2.5
2	2.3
3	2.15
4	2.05
5	2.00

Considering the four test beams the minimum load factor is 2.06 and thus the above requirement is just satisfied (2.05 for four components). It would be more reasonable, due to the large defect in beam 1, to assess the performance of the beams on tests 2, 3, and 4. In this case the minimum load factor is 2.77. It is also apparent that the use of a two species web does not significantly affect the performance of the beams in terms of strength and serviceability. The code also states that the design load multiplied by the appropriate load factor should be sustained for a period of fifteen minutes without failure.

A testing procedure for joints involving metal fixings such as nails, screws and bolts has been recommended by Sunley and Brock. (32). The number of tests should be sufficient for a statistical analysis to be made, say twenty to thirty. As with the derivation of basic stresses the working load is obtained from the expression

$$W = \frac{\bar{x} - 2.335\,\sigma}{F.} \qquad\qquad 15(ii)$$

where W is the working load
       $\bar{x}$ is the mean ultimate load
       σ is the standard deviation
       F is the load factor

The standard deviation is multiplied by a factor of 2.33 for one per cent exclusion. The value to be adopted for the load factor F has been considered in Chapter 9. Sunley and Brock point out that the specific gravity of the material of the specimens and its standard deviation should be checked against known values for the species as a whole. These values can be obtained from reference 9. If the joints are made with above average material, allowance should be made for this in the evaluation of the working load. Where the coefficient of variation (ratio of standard deviation to mean value) is below 0.1 the actual values should only be used if there is an acceptable reason for the low variability.

With metal fixings failure generally occurs due to stress concentrations set up in the timber and thus the coefficient of variation is likely to be in excess of 0.1. A value of less than 0.1 could possibly occur if the joint was designed in such a manner that the fixing itself failed rather than the timber.

16   And what of timber's future?

It has been suggested that all timber members in the future will be reconstituted from a vat of wood pulp, from which material will be extruded into whatever form is required. In this way timber becomes freed completely from its natural limitations of dimension, and can be formed into all manner of plastic shapes, which timber in its natural form could never accommodate. Already there is movement towards the elimination of timber waste by using smaller sections of wood glued together to form the appropriately larger sized members. Even planar waste is used to make chipboard. These processes are already overcoming timber's dimensional restrictions, and extending its strength characteristics. In fact the results of the first moves in this development are already a part of our daily lives—plywood, particle board, laminated timber structural members. Nevertheless, although we are bound to see a continuation of this trend as timber becomes a more valuable material and glue technology advances, it is likely that simple structural members will still continue in their simple form. After all, they display such good natural properties that for such straightforward requirements, it is difficult to improve on Nature.

There is no doubt that rather than timber being an old fashioned material, it is one with great relevance to the building of today and tomorrow. There is more acceptance now of its intrinsic merits than has been the case for several hundred years—in this country at any rate. And many of the misconceptions that attach to timber as a structural material have now been destroyed. We can expect in the next few years to see further advances in the chemical treatments of timber to protect it even more effectively from fire, rot and infestation. We may well see a movement away from the more or less general and thoughtless use of gloss painted finishes on joinery timbers and an acceptance of the more natural preservative stains which are used extensively in Scandinavia. We certainly can expect to see more timber industrialised systems and prefabricated components. The special characteristics of workability, lightness and strength make timber particularly applicable to industrial manipulation, easy site handling and speedy erection—and, where desired, demounting.

While the earth's resources in timber are vast, we can expect a rise in the 'valuableness' of timber and a consequent increase in price. There is no indication, however, that it will be surplanted for a considerable time, as the cheapest spanning building material.

There will, however, be more restrictions placed on the felling of our forests in attempts to preserve the shreds of our environment, but this already is being offset in those countries, such as Britain, who have an afforestation policy. Generally the attempts to restock the supplies of timber have been left very late, but it is to be hoped that now, in this time of sensitivity about the environment, really large sums of money will be spent in planting this most valuable long term crop.

Finally maybe we should consider what building is all about in the

latter part of the twentieth century. Once man built for as long a period
as possible. He erected a monument to himself that hopefully might
last forever. He often did not succeed in building his own immortality,
but it rarely entered his head that what he was building might only have
a limited relevance—and be a positive embarassment to future generations. Compare Paxton's Crystal Palace design with that of Brunel.
Paxton designed a building for precisely the requirements in his brief.
It had to be erected quickly—last for six months—and then be demounted.
The fact that his building demounted so perfectly with a large proportion of re-usable assemblies allowed the further benefit of its re-erec-
tion for another period in an enlarged form.

Many of our buildings today have only a limited relevance. Just compare the heritage of obsolete school and hospital buildings with the
buildings that suit the requirements of today. In short, there is a strong
argument for buildings for certain purposes being designed for a limited life—rather like the washing machine and the motor car. But unlike
these articles, the building should be easily removed and, perhaps,
reformed in a different form. This is the area in which light industrialised system buildings have particular application. No doubt timber has
a significant part to play in this sector of building.

Development in building technology will certainly move towards lighter,
more quickly assembled and more readily demounted buildings. The
requirement for new building becomes more urgent and changes more
quickly, year by year. Much development must be connected at least
in part with timber. No doubt other materials such as plastics will be
used, but timber combines good physical properties with beauty, and
will be hard to pass over, particularly now that many of its natural limitations are being overcome.

The oldest building material will still be a valid building material for
many years to come.

Glossary

BASIC STRESS	The stress which can safely be permanently sustained by timber containing no strength reducing characteristics.
GRADE STRESS	The stress which can safely be permanently sustained by timber of a particular grade.
DRY STRESS	A stress applicable to timber having a moisture content not exceeding 18 per cent.
GREEN STRESS	A stress applicable to a timber having a moisture content exceeding 18 per cent.
PERMISSIBLE STRESS	A stress which can safely be sustained by a structural component under the particular condition of service and loading.
STRENGTH RATIO	The ratio of grade stress to basic stress.
BARK	the external group of tissue from the cambium outwards of a woody stem.
BORDERED PIT	a pit in a trachied or other cell of secondary xylem having a distinct rim of the cell wall overarching the pit membrane.
CAMBIUM	a layer one or two cells thick which is capable of active cell division.
CELL	a structural and physiological unit composing living organisms, in which take place the majority of complicated reactions characteristic of life.
CELLULOSE	a complex carbohydrate occurring in the cell walls of the majority of plants.
CONIFER	cone bearing tree.
LIGNIN	an organic substance or group of substances impregnating the cellulose framework of certain plant cell walls.
PIT	minute thin area of a cell wall.
PITH	tissue occupying the central portion of a stem.
TISSUE	a group of cells of similar structure.
TRACHIED	an elongated tapering xylem cell adapted for condition and support.
WOOD	a piece of tree, generally the xylem.
XYLEM	a plant tissue consisting of trachieds and other cells.

References

1 Lavers Gwendoline M. The strength properties of timbers, Ministry of Technology Forest Products Research, Bulletin No. 50, London H.M.S.O. 1969.

2 Moroney M. J., Facts from figures, Penguin Books.

3 British Standards Institution, The structural use of timber Part 2. metric units, CP 112: 1971.

4 Booth L. G. and Reece P. O. A commentary on the British Standard Code of Practice CP112: 1967, E. & F. N. Spon Ltd. 1967.

5 The Swedish Timber Council. The properties of Swedish Redwood and Whitewood, July 1971.

6 de Bruyne N. A. et al. The theory and practice of gluing with synthetic resins, Large, Maxwell and Springer.

7 Timber Drying. Timber Research and Development Association in conjunction with the Timber Drying Association, July 1969.

8 Silvester F. D. Visual stress grading of timber, Timber Research and Development Association 1969.

9 The Swedish Timber Council, Swedish Redwood and Whitewood, The Sawn Timber and Products, July 1971.

10 Council of Forest Industries of British Columbia (CFI). 44 mm Hem-Fir, Canadian Metric Timber for the British Builder.

11 Langlands I. The mechanical properties of South Australian plantation-grown Pinus Radiata Pamph. Coun. scient. ind. Res. Aust. No. 87, 1938.

12 Curry W. T. Mechanical stress grading of timber, Timberlab paper No. 18—1969, Forest Products Research Laboratory*, Princes Risborough, Bucks.

13 Curry W. T. Interim stress values for machine stress-graded European Redwood and Whitewood, Timberlab paper No. 21—1969, Forest Products Research Laboratory, Princes Risborough, Bucks.

14 The properties of Swedish Redwood and Whitewood, Swedish Timber Council, Swedish House, Trinity Square, London EC3.

15 44 mm Hem-Fir, Council of Forest Industries of British Columbia, Templar House, 81-87 High Holborn, London WC1X 6L5.

16 The structural use of hardwoods, Timber Research and Development Association, Hughenden Valley, High Wycombe, Bucks.

17 Burgess H. J. and Peek J. D. Span charts for solid timber beams, Timber Research and Development Association 1968.

---

* Now called the Princes Risborough Laboratory of the Building Research Establishment (see Acknowledgements)

18   Fir plywood design fundamentals and physical properties, PMBC, United Kingdom Office, Templar House, 81-87 High Holborn, London WC1X 6L5.

19   Burgess H. J. Introduction to the design of ply-web beams, TRADA information bulletin E/1B/24.

20   Timoshenko S. P. and Gere J. M. Mechanics of materials, Van Nostrand Reinhold Co. 1972.

21   Fir plywood web beam design, CFI Templar House, 81-87 High Holborn, London WC1X 6L5.

22   Tottenham H. The effective width of plywood flanges in stressed skin construction, Timber Research and Development Association E/RR/3, March 1958.

23   Fir plywood stressed skin panels, PMBC, Templar House, 81-87 High Holborn, London WC1X 6L5.

24   Plywood folded plate design, PMBC, Templar House, 81-87 High Holborn, London WC1X 6L5.

25   Stern E. G. Research on jointing of timber framing in the United States of America Modern Timber Joints, Timber Research and Development Association. 1968.

26   Jansson I. Nailed joints with plywood, Modern Timber Joints, Timber Research and Development Association. 1968.

27   Ministry of Technology, Forest Products Research Laboratory, Strength tests on structural timber joints made with 20-gauge Gang-Nails. April 1966.

28   Macandrews and Forbes Ltd., Design manual for timber connector construction.

29   Wardle T. M. Glued scarf and finger joints for structural timber, Timber Research and Development Association. 1967.

30   Ghugg W. A. Glulam, Ernest Benn Ltd., 1964.

31   Enger K. and Kolb H., Investigations on the ageing of glues for supporting structural wooden elements, Otto-Graf-Institute, Technical University, Stuttgart.

32   Sunley T. G. and Brock G. R. Joints in timber, methods of test and the determination of working loads, Modern Timber Joints, Timber research and development association 1968.

Further reading

Robbins, Weier and Stockway, Botany—An Introduction to Plant Science, John Wiley and Sons Inc., New York.

Edlin H. L., The Forestry Commission, A Review of Progress to 1971, HM Stationery Office, 1971.

Edlin H. L., Timber! Your Growing Investment, Forestry Commission, HM Stationery Office, 1969.

Taylor G. R., The Doomsday Book, Thames and Hudson 1970.

Parsons W. B., Engineers and Engineering in the Renaissance, MIT Press, 1968.

Beckett D., Bridges, (Great Buildings of the World), Hamlym 1969.

Condit C. W., American Building Art—The Nineteenth Century, Oxford University Press, New York 1960.

National Building Studies—Special Report No 36, Prefabrication—A History of its Development in Great Britain, HM Stationery Office, 1965.

The Royal Institute of British Architects, The Industrialisation of Building, RIBA 1965.

National Building Agency—NBA Metric Component File, NBA 1971.

Hall F. A., Glued Laminated Timber Structures, Rainham Timber Engineering Co Ltd.

Sunley J. G., Timber, Building and Technological Advance, Timberlab Paper No 9, Building Research Establishment, Princes Risborough Laboratory, 1969.

Council of Forest Industries of British Columbia, Canadian Fir Plywood Manual.

Timber Trade Federation of the UK, Plywood for Building and Construction.

Ashton L. A., Fire and Timber in Modern Building Design, Timber Research and Development Association, 1970.

TRADA, Timber and Fire Protection, Timber Research and Development Association, 1953.

Ministry of Technology, Symposium No 3, Fire and Structural use of Timber in Building, 1967.

Technical Note No 25, Preservative Treatment of External Joinery Timber, Building Research Establishment, Princes Risborough Laboratory.

Cockcroft R., Timberlab Paper No 46, Timber Preservatives and Methods of Treatment, Building Research Establishment, Princes Risborough Laboratory, 1971.

White M. G., Timber lab Paper No 33, The Inspection and Treatment of Houses for Damage by Wood-boring Insects, Building Research Establishment, Princes Risborough Laboratory, 1970.